PRA

The Brigade

"Remarkable. . . . *The Brigade* [is] an illuminating addition to the annals of World War II."
—*New York Times Book Review*

"Compelling. A great cloak-and-dagger story. . . . [Blum] raises moral questions that transcend any single time, place, or side."
—David Hinckley, *New York Daily News*

"Howard Blum's remarkably written and documented *The Brigade* is more than a story about memory, justice, and vengeance; it is a moving evocation of events and people whose tears and dreams will continue to haunt humanity for centuries to come."
—Elie Wiesel, Nobel Laureate and author of *Night*

"A powerful wartime saga that is also a meditation on morality. One of the most moving accounts of the war to appear in many decades."
—*Washington Post*

"*The Brigade* is a gripping historic tale of a war's end and a nation's beginning. Howard Blum tells this true story with the narrative pace of a novel."
—Dan Rather

"*The Brigade* is a riveting, eye-opening read. Blum deftly depicts the heroic actions of these men whose battle against German machine-gun nests is merely a prelude to an even more poignant struggle against despair and their own bloodlust—and who, in the midst of hell on earth, find the rarest of all things: redemption."
—Dennis Lehane, author of *Mystic River*

"An action-packed, real-life drama. . . . Military historians, fans of war stories, and lovers of Judaica all will be pleased."

—*Kirkus Reviews*

"In *The Brigade*, Howard Blum brings alive the dramatic final chapter of World War II. Fascinating and evocative, *The Brigade* takes us into the murky depths of a post-war world that have been too little explored. Howard Blum has given us a compelling story of war and vengeance, rich with political intrigue and personal drama."

—Jay Winik, author of *April 1865: The Month That Saved America*

"A gripping story, superbly told as only Howard Blum can tell it. . . . Full of derring-do and intrigue, of pain and courage, nobility and humor—and all true. Read it; you will love it."

—Irving Greenberg, president, Jewish Life Network, chairman, United States Holocaust Memorial Council

"Blum . . . presents the material masterfully, building suspense and carefully documenting all the action." —*Publishers Weekly*

Jenny Blum

About the Author

HOWARD BLUM is the bestselling author of *I Pledge Allegiance, Out There, Wanted! The Search for Nazis in America, Gangland, The Gold of Exodus,* and the novel *Wishful Thinking*. As a reporter for the *New York Times*, Howard Blum was twice nominated for the Pulitzer Prize. Currently, he is contributing editor to *Vanity Fair*. Mr. Blum lives with his wife and three children in Connecticut.

THE BRIGADE

An Epic Story of Vengeance, Salvation, and World War II

HOWARD BLUM

Perennial
An Imprint of HarperCollins*Publishers*

Once again for Jenny.
Always with love.

A portion of this book was published in slightly altered form in *Vanity Fair.*

A hardcover edition of this book was published in 2001 by HarperCollins Publishers.

First Perennial edition published 2002.

Designed by Joseph Rutt

The Library of Congress has catalogued the hardcover edition as follows:
Blum, Howard.
The brigade : an epic story of vengeance, salvation, and World War II /
by Howard Blum.— 1st ed.
p. cm.
ISBN 0-06-019486-3 (alk. paper)
1. Great Britain—Army—Jewish Brigade—Biography. 2. World War, 1939–1945—
Personal narratives, Jewish. 3. World War, 1939–1945—Participation, Jewish.
4. Jewish soldiers—Palestine—Biography. 5. Carmi, Israel. 6. Pinchuk, Arie.
7. Peltz, Johanan. I. Title.
D810.J4 B529 2001
940.54'125694—dc21 2001024324

ISBN 0-06-093283-X (pbk.)

02 03 04 05 06 ❖/RRD 10 9 8 7 6 5 4 3 2 1

Then out spoke brave Horatius, the Captain of the Gate, "To every man upon this earth, death cometh soon or late, and how can man die better than facing fearful odds, for the ashes of his fathers and the temples of his gods."

—MACAULAY

A NOTE TO THE READER

This is a true story.

It is an account of the activities of the Jewish Brigade, an army of five thousand men from Palestine sent by the British to fight in Europe during the last months of the Second World War.

It is also the story of three soldiers—Israel Carmi, Johanan Peltz, and Arie Pinchuk—who volunteered to serve in this force. They were men whose lives were transformed by the times in which they lived; and, more consequentially, they were men whose actions transformed the world around them.

To accomplish these two intertwined narrative ambitions, one thread historical, the other personal, I have employed a variety of sources: 114 separate interviews; government archives in Britain, Israel, and the United States; and a small mountain of reference books, published in both English and Hebrew. A detailed chapter-by-chapter sourcing follows at the end of this book.

However, since *The Brigade* is so much a story about these three men—how they lived their lives, and how the resolutions of their own internal conflicts had an effect on larger events—it is important for the reader to know from the outset how I can write about them with such knowledge and authority.

I had a reporter's most valuable substantiating resource: eyewitness testimony. Not only did all three men sit for a series of lengthy interviews—I have their words on tape, on video, and on the pages of the yellow legal pads I filled at each of our sessions—but they also made available their own unpublished memoirs. These were thoughtful and detailed works (Carmi's went on for 239 single-spaced pages, while Peltz's was more than twice that), exciting to read and remarkably candid.

In the "Note on Sources" that appears at the end of this book I will document chapter-by-chapter how these memoirs, as well as Leah Pinchuk Zeiger's own reminiscences, were unique and invaluable sources. But for now it should be sufficient to say that to a significant degree these unpublished autobiographical works were a resource that enabled me to report accurately what people were thinking, feeling, and saying sixty years ago. Without them, I could not have written this true story in such a fashion.

PROLOGUE

"THIS IS THE TWILIGHT"

September 1944

It was only a hundred days since the invasion at Normandy, but as the first cool nights of fall arrived, the Allied command was confident the war in Europe would soon be over. Paris had been liberated. The lights were back on in London. The Russians were fighting in Tallinn, Riga, and the streets of Warsaw. And the Allied troops—the largest army in the history of the world—were moving rapidly to cross the Rhine. There would be more combat, and more deaths, but by September 1944, the outcome of the war was no longer in doubt.

Meeting with correspondents as his troops pushed through France, Gen. Dwight D. Eisenhower, the Supreme Allied Commander, exuberantly declared that "Germany's military situation was hopeless." Many in the Nazi high command had also come to a similar conclusion. "There comes an evening in every campaign," Lt. Gen. Frederick Heim, commander of the captured garrison at Boulogne and veteran of two previous encirclements at Kiev and Stalingrad, conceded on that same day in September. "This is the twilight."

It was in this "twilight" that Prime Minister Winston Churchill realized he had to act on the petitions from the Jewish Agency in Palestine, or soon it would be too late. Since the start of the war the Jews in this British Mandate territory had been lobbying for the opportunity to fight the Nazis as a separate unit in His Majesty's army. For five years, the British had found excuses—shortages of equipment, training problems, recruitment difficulties—to justify their long reviews and ultimate refusals. The truth, however, was bluntly expressed by the British secretary of state for war. "We have not asked the Jews for these units," he wrote to the colonial secretary, "and we do not want them for our war purposes. On the contrary, the War Office have always made it quite clear that they would regard them as an embarrassment rather than a help."

There was reason for this animosity: the British government knew they were dealing with men who had fought against them in the past, and very likely would be their enemies in the future.

In May 1939, the British had issued a legislative proposal in the form of a "white paper" on the Palestine territory. For the Crown, the White Paper was a pragmatic act of statesmanship; a world war was inevitable and they would need the support of the Arab nations dispersed across the Empire's eastern flank. For the Yishuv, as the community of Jews in Palestine was known, it was a devastating piece of politics, a shameless renunciation of past promises that had guaranteed the establishment of a Jewish homeland. The new policy severely restricted the sale of land by Arabs to Jews. Jewish immigration was reduced to a trickle just as Hitler's intentions were becoming ominous. And of even greater concern to a community still recovering from the violence of the 1936 Arab Revolt, the White Paper announced that an independent Palestinian state with an Arab majority would be established at the end of five years.

The Jews were furious and frightened. In this angry and combative time, amid days of riots and general strikes, the Haganah, the Yishuv's officially outlawed but unofficially tolerated underground army, abandoned its previous restraint against the Mandate authorities. Its fighters attacked government railways and offices, and held off British troops as boatloads of illegal immigrants frantically waded ashore to find cover in the sandy dunes that ringed Tel Aviv.

However, when the war in Europe broke out in September the Yishuv drew back from its aggressive opposition to the British—a bit. David Ben-Gurion, the head of the Jewish Agency in Palestine, hoping simultaneously to reassure both his people and His Majesty's government, articulated this nuanced attitude with Solomonic precision when he said: "We will fight the White Paper as if there is no war, and fight the war as if there is no White Paper."

The British were not persuaded. When the Jewish Agency quickly volunteered to recruit its own force to serve alongside the king's troops against the Nazis, the offer was rejected with corresponding alacrity. "Your public spiritied assurances are welcome and will be kept in mind," Prime Minister Neville Chamberlain tactfully wrote back.

Instead, the government decided it would allow Jews to join the Buffs, the Royal East Kent Infantry Regiment already stationed in Palestine. As a further restriction, it insisted that Jews and Arabs be recruited in equal numbers. But this parity, however prudent a diplomatic policy, was never a reality. By August 1942, when the Buffs were merged into a new colonial force, the three battalions of the Palestine Regiment, there were more than three times as many Jewish volunteers in the British army as Arabs.

Yet as the plans to enter Berlin were being drawn, as the first eyewitness reports about the systematic extent of Nazi atrocities were beginning to be published, Churchill decided he had had enough of "the usual silly objections." He sent a personal telegram to President Roosevelt arguing that "the Jews . . . of all other races have the right to strike at the Germans as a recognizable body." Five days later the president replied: "I perceive no objection. . . ."

On the evening of September 19, 1944, when Rosh Hashanah, the celebration of the Jewish New Year, began, the British War Office published an official proclamation: "His Majesty's government have decided that a Jewish Brigade should be formed to take part in active operations. The Infantry Brigade will be based on the Jewish Battalions of the Palestine Regiment. The necessary concentration for training is now taking place before dispatch to a theatre of war. . . ."

The next day, the *New York Times* reported the announcement on page 12, and dismissed the Brigade as a "token." "Its sponsors," it was cautiously noted in the brief article, "hope the group will be available before the final battle for the destruction of Nazism."

The editorials in the British press, while supportive, were also similarly guarded. The *Times* of London saw the formation of a Jewish Brigade as "symbolic recognition." The *Manchester Guardian* complained that the announcement was "five years late." *The New Statesman and Nation* offered grudging praise: "Late though it comes, we welcome this victory over anti-Semitic prejudices."

It was generally agreed that the decision to send a Jewish Brigade—a mere five thousand men—to fight in Europe was of no military significance. The war, it was believed, would be over by Christmas or, at the latest, New Year's. The formation of a Jewish fighting force was simply a gesture, pure politics. It was only a footnote to larger events, one that would have no influence on the course of the war, or the shape of the peace. And certainly one that would have no effect on the flow of history.

PART I

AT SEA

November 1944

ONE

The troops were singing. The Hebrew songs had broken out spontaneously when the trucks approached the docks in Alexandria and the men saw the cargo ships that would take them across the Mediterranean to Italy. As they marched up the gangplanks, their voices grew louder and more spirited. It was a bright Tuesday morning, the final day of October 1944, and at last the Jews from Palestine were going to war.

For the past three years, the volunteers in the three battalions that made up the British Palestine Infantry Regiment had spent their days doing monotonous guard duty in North Africa, chasing after goats stolen by mischievous Arab youths, and training in the hills north of Tel Aviv with outdated weapons. A world away the war in Europe had raged on. But just five weeks ago the three battalions had been ordered to Burg-el-Arab, a flat brushless stretch of desert between El Alamein and Alexandria, and had been swiftly re-formed and outfitted into the combat-ready Jewish Brigade Group.

The historical significance of these soldiers preparing to go off to Europe with a golden Star of David on their blue shoulder patches was appreciated by the Brigade's new British commander. Brig. Ernest Frank Benjamin announced to his officers that "this is the first official Jewish fighting force since the fall of Judea to the Roman legions."

But as the boats left the harbor, the men standing on the quarterdeck watched the Egyptian shore disappear and were excited by another knowledge. An army of Jews was finally on its way to confront an enemy that had set out to annihilate their people.

Sgt. Israel Carmi did not go on deck with the others as the SS *Stafford* went to sea. He remained on guard by the bunks.

He had been ordered by the Haganah to smuggle two men to Europe along with the troops. Days ago he had stolen two uniforms, and his wife had tailored them at the kibbutz so that the impostors could march up the gangplank unnoticed with the rest of the soldiers.

But Carmi knew the two would never be able to fool a British officer. So he stayed with them as they lay on their bunks. He would try to intervene before anyone could become suspicious. He was prepared, though, to do whatever was necessary to ensure that the two stowaways arrived in Europe. Carmi was a sergeant in the British army, but his allegiance was to the Haganah and the land, the *eretz*, he was leaving behind.

That night, Capt. Johanan Peltz could not sleep. He went up to the deck and walked over to the rail. The moon was high in the sky and illuminated the sea with a silver sheen. The shoreline had receded from sight, and he was full of anticipation. He had spent seven years in this primitive, overbearingly hot part of the world, and he was glad to be leaving.

It felt blasphemous to be happy in these grim times, but he could not help being filled with a sense of joy as he imagined returning to Zabiec, his family's estate in Poland, once the war was over. In his mind it was all very clear. He would ride in a carriage through the allée of chestnut trees to the front door of the big brick house. It would be the dinner hour, everyone certain to be home. He would walk up the wide stone steps and when the maid answered the door he would tell her not to announce him. Then he would stride across the checkerboard floor of the main hall, tall and erect in his British officer's uniform, and into the dining room, while his parents and grandfather, proud and elated, rushed from the table to greet the returning hero.

Lt. Arie Pinchuk was in the radio room playing bridge. He, too, could not sleep. His stomach bubbled nervously. He was an officer in the Jewish Brigade, but he was traveling to Europe on his own private, and very vital, mission.

He could not concentrate on the cards, either. He was thinking about what he would need to do, all the obstacles he would face. And from the depths of his own internal hell, he was silently demanding, *Mama, Papa, Leah, what has become of you?*

It was nearly three A.M. when Pinchuk returned to his bunk, and by then the *Stafford* was rolling violently. The ship had headed into a storm. Waves crashed against the bow, and the swells surged over the deck. The winds were shearing. Yet the convoy continued on. All the men could do was hope the rough weather would soon improve.

TWO

The storm grew worse. Rain poured down throughout the morning, and the wind continued to blow hard. The *Stafford* bounced on high, dark waves, its stern lifting up in the water as its bow dropped. Swells slammed against the boat, and the big sea swept over the deck. Across from the pilot house, crests of water surged repeatedly over a hill of wooden crates and the ropes that bound them finally gave way. The boxes slid about the deck, smashing into one another and careening into the rough, rolling sea.

The men below were tossed about helplessly, too. Nevertheless, when his relief came to assume the watch over the two stowaways, Carmi was eager to try to move about. He wanted to escape the stuffy, crowded hold deck with its tiers of bunks, and the engines vibrating ominously beneath him as the ship struggled forward.

He worked his way along a narrow, heaving corridor to the bulkhead door and started to the top. But guards were posted to prevent any reckless soldiers from going on deck. Instead, Carmi was directed to the mess.

He was in no mood for food and, apparently, neither were many of the men. The mess was nearly empty. But just as he was about to leave, he noticed an officer sitting alone, on a bench attached to a long gray table across the room. The man's back was rigidly straight, his blue eyes staring ahead as though fixed on a distant object, his hands clasped together on the tabletop.

Nearly six years had passed since they had served together in the Palestinian Police, but Carmi recognized the soldier immediately. Of course, he had heard the stories. He knew the man had been declared one of the *porshim*, judged unreliable by the Haganah. The commanders of the underground had even tried to prevent him from becoming an officer in the British army. But this was of little significance to Carmi. When he saw Johanan Peltz sitting across the room, he felt only a rush of affection for an old friend.

The men had first met in the winter of 1937, in the riding ring of the Police School near Mount Scopus in Jerusalem. The new recruits were choosing mounts. Peltz was inspecting a tall sturdy chestnut, at least eighteen hands, with a white star on its forehead and a matching white sock on its right foreleg. The horse had a disagreeable disposition, snarling menacingly as Peltz moved to pat its flank.

When the Haganah had begun looking for riders to serve in the colonial force, Carmi had matter-of-factly volunteered, mounting a horse for the first time. Despite his diligence, a year later he was still a bouncing thick-bodied horseman. "Lot of horse," Carmi warned, trying to be helpful.

Peltz interpreted the veteran's remark as a challenge. "I'm a lot of rider," he said, and quickly climbed on the horse. Once he was in the ring, he put on a calculated and masterful performance. It ended with his reaching down to the ground while on a full gallop to grab one of the blue kolpaks the Palestinian Police wore as hats.

"She'll do," he said as he jumped off.

But there was a problem. The British cavalry sergeant insisted that if Peltz wanted this animal as his regular mount, he would have to name it. Peltz studied the horse, but nothing came to mind. The sergeant watched him, and grew impatient.

"Something the matter with you, boy? Don't have a brain?"

Peltz bristled.

"If you can't think of a name," Carmi said, "call her 'Anonymous.' "

Peltz immediately agreed. That night the two men ate dinner together, and, as they stayed up late talking afterward, a friendship was begun. Carmi appreciated Peltz's swaggering confidence, his eagerness to accept a challenge. Peltz respected the Haganah veteran's uncompromising determination to fight for what he believed in. They spent a lot of time in each other's company that cold winter in Jerusalem. But in the spring Carmi was stationed in the north and Peltz was given command of an isolated police post on the southern end of the Dead Sea.

They had not seen each other since then. Yet as Carmi leaned on a bulkhead to steady himself against the roll of the ship, he looked across the mess and reconsidered disturbing his old friend. Peltz seemed in a trance. Perhaps their reunion should wait for another day. What, Carmi wondered, could Peltz be thinking? He decided there was only one way to find out.

"How's Anonynous?"

Surprised, Peltz looked up and saw Israel Carmi standing above him.

"Sold," said Peltz. "Made a good profit. Besides, I can't ride anymore. I took an Arab bullet in the knee and now the whole leg's no good. Can't get a grip on a horse."

"I heard," said Carmi, letting Peltz know that he had heard *everything*: the ambush at Sdom, as well as the dispute with the underground commanders that had followed. And that while Carmi was Haganah, he was also still his friend.

"Sit, Israelik," Peltz said gratefully.

Carmi sat down across the table, and told Peltz that he had been watching him. "I almost walked away. You looked deep in your own world."

"I was."

"Well?"

Peltz thought for a moment, and then decided to share what had been going through his mind. When Peltz was six, a new riding instructor came to live at Zabiec. His name was Tony Power and he was an Oxford-educated English army captain who had originally been sent to Poland as a liaison to the cavalry. Power taught Johanan until the boy went to high school, and Peltz idolized him. In their years together Peltz not only learned to speak a clipped and precise King's English, but he also tried to mimic the casual, yet confident superiority that was the essence of Power's Englishness. And now Peltz told Carmi a story he had first heard from Power.

The king of England, Power had reprimanded the young, restless Peltz, was able despite the demands of nature to sit rigidly in the saddle while the troops passed in endless parade. " 'Know how His Majesty managed to combat the urge to pee?' " Peltz quoted. " 'He made up his mind he didn't need to.' "

"And you were thinking about this for a reason, Johanan?" Carmi asked.

"Seasickness," Peltz said. "Everyone's sick to their stomachs. But not me. I make up my mind not to be sick, I'm not sick."

Carmi considered this, then he said, "I can last as long as you can."

"We'll see," said Peltz.

As the boat surfed high and crashed down with each new wave, the two men sat opposite one another and talked. They did not discuss the war. They spoke about other matters. A sly acquaintance from their police days who furtively juggled two girlfriends, both waitresses at the same café in Jerusalem. A raid Peltz had led on Madame Muneira's infamous "Kilo Seven," the brothel named for its location directly opposite the "kilometer seven" stone road marker on the road to Motza. How Carmi and his wife were trying to teach Shlomit, their infant daughter, to walk. They talked and talked, hoping to keep their minds occupied with easy, comfortable conversation, rather than on the unsettled feeling that was starting to rise up in the pits of their stomachs.

It was only a small contest, but each man was determined to win.

* * *

Down below, Arie Pinchuk was sick. Violently sick. He had spent most of the afternoon running from his bunk to the latrine. But now he got up and forced himself to tour the pitching ship. He hoped that the exercise would be a palliative. It had to be better than the agony of lying on his bouncing bunk surrounded by the raw odor of engine fuel and clammy sweat.

Yet after wandering around the ship for just a brief time, he needed to sit. He made his way uneasily to the mess. Entering, he saw Peltz across the room and immediately decided to leave.

As a newly commissioned lieutenant, Pinchuk had been put in charge of a platoon that was assigned to Peltz for training. Peltz had been relentless. Even the British at the Sarafand camp outside Tel Aviv had noticed; a Scottish sergeant nicknamed Peltz "Skinner" because he "skinned his lads to the bone." Endless bayonet drills, rifle practice until calluses rose on many of the men's trigger fingers, and arduous field marches, the men in full kit and pounding at a breathless pace through the solid dunes of Rishon Le Zion under an unforgiving sun—Peltz kept his troops busy with a strict, almost cruel diligence. Pinchuk had found it unbearable. Lying down on his hard cot at night, he was aching and totally spent. He grew to detest the autocratic Peltz.

And now when Pinchuk saw his old tormentor, his first instinct was to find some other place to sit. But then he saw the husky man on the bench across from Peltz, and he hesitated.

He had never met Israel Carmi, but he knew about him. He had heard stories about what Carmi had done in the Haganah. And like many of the soldiers, Pinchuk believed that if it had not been for the mutiny Carmi had led when the Second Battalion was stationed in Benghazi, the Brigade might never have been sent to Europe. If the British had not been forced to confront the anger and resolve of the Jewish soldiers, they might very well have kept them permanently out of the war. He was convinced that Carmi was an authentic Jewish hero; and without Carmi's actions, his own mission never would have been possible.

So while Pinchuk was normally a shy, guarded man, he felt he could not walk away from the opportunity to meet Israel Carmi. He went over to the table. To his delight, a moment later Carmi asked him to sit down. And Peltz, always competitive, soon explained the game they were playing.

"What about you, Arie?" he asked. "Want to play?"

THREE

It was an odd competition, and of no apparent consequence. But each man stubbornly refused to lose. And as they struggled to persevere, the three soldiers found they were creating a new, intimate fraternity. Lesser men lay below, moaning and retching as the *Stafford* rolled heavily across the sea, while they alone dared to face up to the storm.

At the same time, Pinchuk was also miserable. He had not wanted to accept Peltz's challenge, but it would have been too humiliating to decline. And now that he was part of the contest, he did not want to give Peltz, so clearly eager to win, the satisfaction of an easy victory. An entirely different sort of pride made it important that he not embarass himself in front of Carmi.

But it was difficult. Pinchuk had to summon up all his discipline, all his concentration, as he battled the effects of the weather. He also had trouble fitting into the flow of conversation. Arie was not by nature a social man. He did not have a repertoire of glib, lighthearted tales. He did not make idle talk. He shared his thoughts. And so when it was his turn to speak, Pinchuk found himself telling a story that his father had often repeated to him. It was a memory that was very much in his mind as the boat brought him closer to Europe.

During the 1914 war, when the front abruptly moved toward the

Bug River, the Cossacks had stumbled upon his birthplace, a village in the western Ukraine called Reflovka. "The whole place," Pinchuk explained with a Yiddish metaphor, "was only as big as a yawn." Still, the Cossacks found great sport in robbing the houses and shops of the Jews.

In the midst of all the commotion, his father had spotted Ivan, a Ukrainian who was close to the family. "He would light the stoves on the Sabbath and close the lights in the synagogue," Pinchuk said. "He made quite a good living from these services, and we treated him as a friend. Well, my father saw him running with a sack of flour he had stolen from a Jewish store. A neighbor saw him, too, and asked, 'Ivan, what are you doing?' And my father saw that Ivan stared at him with a look that could kill, before running away without even bothering to answer.

"It was just a sack of flour, a thing of not much value at all," Pinchuk continued. "Still, my father wanted me to remember this scene because it told him what will happen to all of us from all the 'Ivans,' all our friends and neighbors, if they ever get the chance."

"I worry," said Pinchuk, "that the Ivans have finally gotten their chance."

No one spoke; Pinchuk's frankness was more unsettling than the storm. But Peltz, finally, could not allow the story to go unanswered. The way he looked at things, there were Jews who were victimized, and Jews who fought back. It was just a question of finding the courage to fight. And so he told them about his grandfather.

When a band of marauding Russian soldiers had broken into their home in Poland, his grandfather, a boy of nineteen but already six feet six, had killed two of the intruders with an iron bar.

As a consequence, his grandfather had to flee Poland that night. He made his way to Turkey, where he found a circus that was offering a cash prize for anyone who could defeat Gregory the Invincible in a wrestling match. He won and was invited to join the circus.

For two years, his "big grandfather," as Peltz adoringly called him, traveled with the circus as its strongman. He toured the Middle East straightening horseshoes and bending iron bars with his large hands.

When he finally returned to Poland, he married the daughter of a rich man. Zabiec, with its 24,000 acres of forests, wheat fields, and orchards, came as the bride's dowry.

"See, Arie," Peltz said, "not all of us worry about the Ivans. Maybe they should worry about the Jews."

Pinchuk was less sanguine. Gregory the Invincible was one kind of opponent. The German army was another. As he started to share his misgivings, a lurching feeling rose in his stomach. He needed to escape the room at once.

But before he could, Carmi rose and announced, "It's late. I'd better go." He walked off with great calm.

Moments later, Pinchuk fled the room.

Peltz, enjoying his victory, remained in his seat, his back ramrod straight. Tony Power, he told himself, could not have done better.

The storm raged through the night, and when Peltz returned to his bunk he could not sleep. From the breast pocket of his tunic, he took the letter he had received more than two years earlier from his big grandfather. He had read it so many times that he had memorized the words. But he always found them comforting.

". . . and so my dearest grandson," the letter concluded, "Zabiec is ours again. It is now clear of all debt and I shall be able to restore it to its old glory. . . . We are looking forward to seeing you in Zabiec in one year's time, two years at the latest. I'll have a saddle horse ready for you. . . ."

Pinchuk, too, had a letter. Each night when he tried to sleep it would force its way into his mind. As the *Stafford* plowed through the waves, Pinchuk lay on his rocking bunk, and the opening sentence rose up into his thoughts.

"Our cow is still alive and giving milk," his father had reassured him three years ago in the last letter Pinchuk had ever received from home.

Pinchuk remembered the family's old, sick cow with the red

patches, and how on a snowy night during his last winter in Reflovka he had helped his father push the beast up the three steps to their house. His mother and Leah, his little sister, had complained about the smell. But there was little choice. If the cow remained in the drafty shed, it would freeze to death and then they would be without milk.

Recalling the struggle to push the recalcitrant cow up the steps, he found himself near to laughing out loud. But his thoughts soon turned, as they always did, to his mother, and father, and sister. And his momentary levity vanished. His escape from Reflovka was an act of betrayal.

In time, he closed his eyes and tried to sleep. But before he did, he repeated the silent, redemptive vow he made every night: *Mama, Papa, Leah—I will save you. Whatever it takes, I will save you.*

Carmi was not interested in the past. His concerns were grounded in the present. It was duty, not any sudden queasiness, that had forced him from the mess. It was his turn to watch over Israel Sapir and Misha Notkin, the two stowaways. But he could not disclose this to Peltz or Pinchuk. The Haganah had sworn him to secrecy.

Carmi sat up most of the night with the two men talking and playing cards. But at one point, Carmi's thoughts turned to their mission. The Jewish Agency in Palestine had sent them to Europe to determine whether the rumors of what was happening in the labor camps could possibly be true. There were reports that Jews were being gassed.

"Misha," he said, "the camps. Do you really think it can be that bad? The way we've heard?"

Carmi was not a religious man, but he believed in God. This faith had persuaded him that the evil could not be as pervasive, as institutionalized, as the information making its way to Palestine had insisted.

"We'll soon find out," Misha told him. "Soon we'll see with our own eyes."

* * *

The next day the weather improved. The winds stopped and the sea grew calm.

There were units from two batallions—more than one thousand men—on the SS *Stafford.* The route across the Mediterranean was deliberately circuitous; the frequent changes in course were a precaution to evade any prowling German submarines. The men were at sea for three more days.

We are looking forward to seeing you in Zabiec . . .

The cow is still alive and giving milk.

Soon we'll see with our own eyes.

In this way, "the first official Jewish fighting force since the fall of Judea to the Roman legions" went off to war.

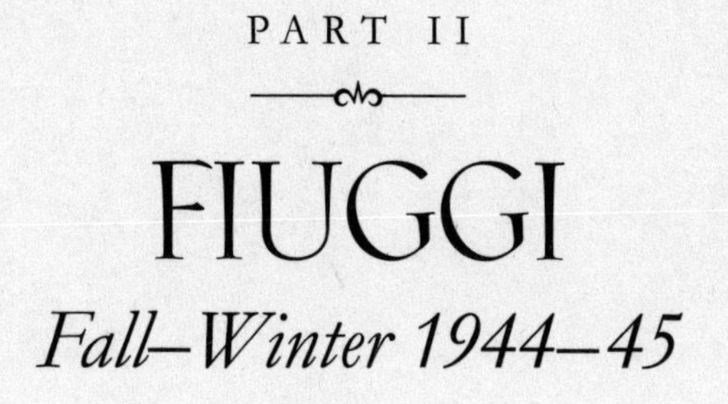

PART II

FIUGGI

Fall–Winter 1944–45

FOUR

In winter a cold, dispiriting rain habitually wrapped itself around the Italian spa of Fiuggi. The mulberry trees lining the streets were bare. The cobblestone squares were deserted. The resort hotels were long shuttered. But in the second week of November 1944, as a punishing rain fell, soldiers crowded into the town.

After disembarking in Taranto, the Jewish Brigade, now part of the British Eighth Army, had traveled north in a long convoy. When they reached Fiuggi, they were ordered to stop and establish a base.

Within days the small off-season resort, about fifty miles southeast of liberated Rome, became a military encampment. Thousands of men were billeted in the town's two fraying grand hotels. Rows of cots were lined up in gilded suites heavy with the smell of fine old dust. Mess was served beneath a huge crystal chandelier in a grand salon whose walls were decorated with frolicking *puti*. Morning calisthenics often were performed in a parquet-floored ballroom; the room was designed for candlelit summer galas, but now only a thin winter's light came through its drafty Palladian windows.

A bottling factory for the town's celebrated spring water became the Brigade's headquarters. Each day the staff officers would make their way to their desks through corridors lined with thousands of empty pale green bottles. At night they would return to their hotel and pass the evening sipping the local weak marsala and playing bridge in a room decorated with fussily upholstered nineteenth-century sofas and armchairs.

That winter the Italian front was quiet. The enemy had pulled back toward the Austrian border and dug in. The twenty-three German divisions had formed a meandering defensive line strung across the hills, ridges, and riverbanks south of Bologna. If the Nazis did not surrender when the weather broke, then they would have to be destroyed. It would be the last great Allied offensive of the Italian campaign. But now it was the long winter and the Jewish Brigade waited impatiently in Fiuggi, preparing for the battles they hoped would come.

The training area was a two-hour march from the village, high up in the hills. On the first day, as Carmi led his platoon along a steep twisting trail up into the alpine ridges, it began to snow. For many of the men—the sabras born in Palestine, those from Africa or Yemen—the flurry of wet white flakes was nearly as astonishing as manna from Heaven. For Carmi, the icy falling snow was a reminder of a former life, and the grim world he had left behind.

A decade earlier he had arrived in Palestine a cautious, German-speaking seventeen-year-old from a wintry Danzig. In the exotic warmth of his new home, he quickly sloughed off his European name of Weinman. Determined to speak only Hebrew, eager to erase the imprint of the past and reinvent himself as a pioneer, he chose the name Carmi and was welcomed into the farming kibbutz of Giv'at Hashlosha.

At the kibbutz, he worked long days under a brittle sun in orchards surrounded by yellow-flowered acacias, their smell wonderfully intense. His pale skin turned a deep, leathery tan. His curly hair sparkled with bright blond streaks. Farmwork filled out his short frame with a solid, compact muscularity.

Confident, proud of his new self-sufficiency, Carmi began to acknowledge his long repressed anger at a European society that had dismissed him as inferior simply because he had been born a Jew. He felt as if he had been reborn in Palestine, and he was immensely grateful to the Yishuv. He anticipated spending a contented life working in the fields, turning the desert into a land of milk and honey.

But the Haganah were shrewd appraisers of untapped talent, and just before Carmi's twentieth birthday they reached out for him. Carmi enlisted without hesitation in the quasi-illegal underground defense force. His initiation was a late-night ceremony in the dimly lit basement of a Tel Aviv high school. With one hand pressed firmly on a revolver and the other on a Bible, Carmi vowed to fight to the death to protect and defend the Yishuv. His days in the fields were over.

On orders from the Haganah, he enlisted in the British-controlled Palestinian Police. It was dangerous duty. In 1936 the Arabs, concerned that the Jewish population in the territory had nearly doubled over the past ten years and, worse, that the British were doing little to discourage this flow of new immigrants, rose up in revolt. Carmi was with the outnumbered squad who fought a day-long battle against the Arab band that had tried to overrun the citrus grove at Kfar Yehezkel. In the Jezreel Valley, he was wounded. And on the northern border he fought the Bedouins hand-to-hand.

Just twenty-one and already a proven veteran, Carmi was selected for service in the celebrated Special Night Squads. Under the eccentric Capt. Orde Wingate, a British colonial officer who wanted his Jewish commandos to attack the Arabs while the bugler blasted a ram's horn, Carmi learned how to take the fight to the enemy's home—and, no less valuable, how to command. For his time with the dashing Wingate, Carmi received the Colonial Police Medal of Valor in an elaborate ceremony at the High Commissioner's Palace in Jerusalem.

The Arab Revolt dragged on for nearly three bitter years, and when it was finally crushed an exhausted Carmi returned to Giv'at Hashlosha. He quickly married Tonka, a pert, blue-eyed woman he had met on the kibbutz. On his wedding night, he promised his bride that he would stay on the farm and grow oranges and melons.

But when the White Paper was issued, there was a new enemy. And the Haganah soon came looking for him.

Carmi was assigned to a group called PUSH. This was a "special operations unit" of the Haganah, and its men were all battle-proven soldiers, all hard cases, and all fiercely loyal. Their mission was to do whatever was necessary to prevent the British security services from arresting the underground's leaders or raiding the covert training camps. It was a violent time, and Carmi was once again on the front lines.

When the war in Europe broke out, Carmi was ordered by the Haganah to enlist in the Buffs. The British wanted to make Carmi an officer. The Haganah, however, thought Carmi could serve the Yishuv more effectively if he worked covertly in the ranks rather than as an officer supporting the British. He became a sergeant.

Carmi had never been entirely comfortable being a British soldier. He could only find two justifications for his service: he would fight the Nazis, and he would acquire the formal military training necessary to defend the Yishuv. But for four frustrating years neither had happened.

When the Brigade had first arrived in Italy, as each successive day they traveled farther north, moving closer to the front, Carmi felt a growing euphoria with the increasing prospect of action. But being bivouacked in Fiuggi for the winter, perhaps even longer, was completely disheartening.

And now, trudging through the Italian snow with his platoon, he found himself asking a familiar, always troubling question: Was this war going to end before an army of Jews had the opportunity to stand up in combat against the Nazis?

FIVE

Carmi still waited impatiently for the Brigade to be called to the front, but most of his grumbling stopped once the training in Fiuggi began.

At last the British were ready to teach the Jews how a modern army makes war. There was instruction in platoon attacks and defense, river crossings, house-to-house and street fighting, mine laying, and signaling and decoding messages.

Carmi, however, was most grateful for the instruction the British gave him and his men on the shooting range. Veteran master sergeants taught the Brigade how to fire Brens and light machine guns, launch mortar shells, and pound away at an enemy with artillery. To Carmi's further amazement, the exercises were conducted with live ammunition; open cartridge boxes were accessible throughout the range. In North Africa, many of the men had marched with broom handles for weapons. And Carmi remembered a time not that long ago when the only way to get weapons and ammunition from the British was to steal them.

The arms thefts had begun during the long, destabilizing summer of 1942. As Rommel's panzers leapt forward across North Africa, the war was suddenly marching closer and closer to Palestine. But it was only after the British began considering withdrawing from the

Middle East and establishing a new line of defense in India that the mood in Palestine turned fatalistic. The leaders of the Jewish Agency and the Haganah decided the time had come to make a plan for the day the Nazis arrived.

It was called *Maoz Haifa*, the Haifa stronghold, and it was an endgame strategy. The Yishuv would gather in the hills and canyons of Haifa, Mount Carmel at their back, and they would fight to the last man. It would be another Masada; the Nazis in the end would be triumphant. Yet there would also be a measure of consolation: the pattern of helpless Jewish surrender would be broken.

But to fight, they had to be armed. It was one thing, the Haganah commanders soberly acknowledged, to mythologize the Yishuv's plight, to encourage the nation to think of themselves as Hebrew Davids standing up to the Aryan Goliaths. However, if rocks were all they had to hurl at this very real and intractable enemy, their courage would be brutally wasted.

As Rommel's troops advanced, Carmi led teams of *rechesh* raiders in an escalating series of arms thefts. They started by furtively walking off with boxes of ammunition and handfuls of rifles from army camps. Then they became bolder, breaking into an ammunition train heading south to Atlit. But even after Montgomery and his Eighth Army had pushed the Germans back across North Africa, the thefts continued. To the Haganah commanders, the opportunity to put away capital for an uncertain future was too tempting.

In Fiuggi, as Carmi lay on his belly in the new fallen Italian snow firing a British Lee-Enfield rifle at a target, he found himself thinking back to just a year ago. At the time he had also been wearing a British uniform. It was Christmas Eve and he had been hiding in the Egyptian sand outside the sprawling Kabrit army base, waiting for a signal.

His back resting against the incline of the wadi, Israel Carmi could hear the soldiers singing. Their strong British voices carried past the rows of tents and out into the desert. "God Rest Ye Merry Gentlemen. Let nothing you dismay . . ." As if to follow their trail, Carmi looked up but saw only a clear night sky filled with stars. "Remember Christ Our Savior . . ." Suddenly angry, he thought, *Your savior, Not mine. Our only salvation is to help ourselves.* ". . . was born on Christmas Day." *And not get caught.*

He listened, trying to keep still. Finally he lifted his head above the ditch. He could see straight across a dirt road to the sentry box by the main gate. The Haganah had told him discipline would be relaxed on the night before Christmas, and he saw only one guard.

Fixing his binoculars to his eyes, he followed the road through the gate and past the mess tent. The arsenal was two tents placed back to back, and they had been deliberately left unlit. He observed with gratitude that there were no guards. Unless, he realized a moment later, the men were inside.

He waited, his body pressed against the cool sand, and listened as the verses of the Christmas carols reached out beyond him, traveling far into the still desert night. And as he waited he grew more and more anxious.

Carmi had stolen arms before, but he had never attempted anything this audacious. He was planning to pillage an arsenal full of Bren machine guns from a British army base manned by thousands of soldiers. And if he succeeded, the guns still had to be transported across hundreds of kilometers of Egyptian desert, past armed border guards on wartime alert, and into Palestine.

But Carmi had been on the line when the Arab Legion, their voices raised in a bloodcurdling battle cry, attacked an isolated kibbutz in the Beit She'an Valley, and there had not been enough rifles to go around. What he would have given for a machine gun that day. So Carmi waited, and tried to be philosophical about the risks.

At last a light shined out from the camp. Carmi scrambled to his feet. After another short flash, he leapt out of the wadi and raced full-speed over the deep hard sand. Looking over his shoulder, he saw three of his men running close behind him, fanned out across the desert.

When they reached the camp, they stopped running and walked ahead casually. There was no pretense of stealth; four men keeping to the shadows, even if they were in British uniforms, might attract curious glances. But there was nothing remarkable about a quartet of loud, tipsy soldiers returning to their tents after the night's celebration. Yet when the four late-night revelers reached the arsenal it was as if they sobered up in an instant. Their theatrically boozy mood vanished, and they were focused on their mission.

The lock was not much of an obstacle. Carmi went in first, his revolver cocked and raised. At the bark of a guard's voice, his plan was to shoot before the challenge could be repeated. But as Carmi walked deeper into the dark tent, the only sound was the careful, measured fall of his footsteps.

He shined his light about, illuminating a large pyramid formed by hundreds of wooden crates.

The men formed a chain. Carmi, meanwhile, opened several boxes until he found the disassembled Brens. He lifted a heavy container filled with machine gun barrels and passed it to the next man.

It was strenuous work. Behind the arsenal, at the edge of a flat plain that was used for marching drills, was a truck that had been driven there by a Jewish corporal who worked in the base motor pool. He was the soldier who had flashed the go-ahead signal. Now the corporal and three of the men were spread out between the tent and the vehicle, the last links in the chain.

It took almost three hours to fill the truck. Carmi saw that the cargo bed was packed to the top of its canvas roof with crates. But as he was going to the driver's cab, he got his first good look at the truck.

There was a swastika on its door, and the insignia of Rommel's Afrika Korps on its hood.

He exploded. What was the likelihood of getting an enemy truck past the British border guards without its cargo being inspected? he demanded.

The corporal from the motor pool began to stammer: this was the only truck he could take without the British noticing. It was not even listed in the official inventory.

Carmi ignored his explanation. In less than an hour reveille would blow. There was no time to load the crates into another vehicle. This truck would have to do.

Before he left, Carmi tied six hand grenades together with a piece of rope, and placed them behind the driver's seat. If the truck broke down in the desert or was attacked, he would blow it up rather than allow the Arabs the opportunity to retrieve the guns.

Finally, Carmi sat at the wheel and two of his men climbed into the cab and they drove off into the thin, predawn darkness. It was only after they were under way that Carmi realized there was a precedent, of sorts, for their Christmas Eve journey. On this night centuries ago, three other travelers bearing gifts had followed the bright stars in the desert sky north toward Bethlehem.

They reached the bridge near the Suez Canal on the Ismailiya-Nitzana Road at first light. A sleepy soldier emerged from the guardhouse, but then snapped alert when he saw the enemy truck.

Carmi, to reassure him, called out in English, and he made sure the man saw his British uniform with the sergeant's stripes on the sleeve.

We're heading up north, to the base in Haifa, Carmi explained. He was hoping that on Christmas morning the soldier would want to go back to bed, and not be too interested in documents.

The soldier, however, asked about the cargo.

Captured enemy weapons, Carmi said. Which, to Carmi's way of looking at things, was the truth.

The soldier nodded and told Carmi to wait. Then he returned to the guardhouse without raising the barrier.

Carmi looked at his men. The longer they waited for the guard to return, the more convinced he became that the guard had gone to get reinforcements.

Carmi ordered his men to get ready. Quickly their guns were below the dashboard, their fingers resting on the triggers.

But the man returned alone, carrying a sack over his shoulder.

"First mail in months arrived yesterday," the soldier explained. "Be a real Christmas present to the lads up in Abu Agila, you could drop this off. Figured since you're heading that way and all."

Carmi discreetly holstered his revolver and took the bag. "Be glad to help."

The soldier raised the gate. "Merry Christmas," he called to the men as they drove by.

That afternoon, however, when they approached the checkpoint at Abu Agila, there were no holiday greetings. One look at the swastikas on the truck and the guards trained their rifles on Carmi and his men.

At the same time, two other soldiers started to undo the canvas in the back of the vehicle. Carmi heard them working the metal latches. When they opened the crates and found the guns, that would be one problem. When they asked to see his travel documents, that would be another.

Carmi knew he would have to act quickly. He could slip the truck into reverse, slam into the soldiers in the rear, and his men would have to do their best at getting off an accurate round at the guards fanned out in front. If he moved without delay, *now,* then at least he would have a chance of controlling the situation.

He was reaching for the gearshift, his mind made up, when he saw the mailbag. He pulled a handful of envelopes from it.

He called out the name on the first envelope. Then the name on the second, then a third.

"That's me!" shouted one of the soldiers as he eagerly grabbed the letter from Carmi's hand. "That's mine!" said another.

Suddenly the soldiers were no longer concerned about procedure, or what was in the boxes in the back of the truck with the swastikas on its doors. It was Christmas morning in the Egyptian desert and there was mail from home. The barrier was raised without further delay.

"I wish you good and happy news from your loved ones," Carmi said. Then he shifted into gear and drove off before the guards reconsidered, and another Christmas miracle would be needed.

By the time he reached Nitzana Udja, the wind had begun to blow. Soon he was driving into a dense dark wall of swirling sand. The

German truck did not have a windshield, so they ripped their shirts and used the cloth to fashion crude bandannas which they tied over their faces.

The storm raged for two hours. Carmi heard the motor grinding, struggling against the waves of sand. But he was afraid to stop. If he did, he doubted he would ever get the sand-clogged engine to turn over. He used his compass to keep the truck pointed north and drove on.

By Be'er Sheva the winds began to subside, and as they entered Gaza the storm had blown itself out. He took an asphalt road toward the Arab city of Lod.

He was driving through the main street of Lod when two British military policemen passed by in the opposite direction. After spotting the swastika on the truck's door, one of the motorcyclists turned back.

"Get ready to open fire at my command," Carmi ordered.

Both soldiers had their revolvers out. One leaned out the side of the truck and took aim.

Carmi pressed his foot hard on the gas pedal, but the motorcyclist gained.

Yet Carmi hesitated. At the sound of gunfire the crowd of Arabs in the marketplace, many undoubtedly armed, would surround the truck. He saw himself in a battle he could not win. The Arabs would be running off with the Brens before the British troops arrived to arrest him and his men.

"Hold your fire," he shouted.

The motorcyclist was nearly even with the truck. The MP shouted at him to pull over. Carmi kept his foot on the accelerator. "That's an order," the MP barked. "Pull over."

Carmi ignored him. The soldier attempted to speed past the truck. But just as he was starting to pull ahead, Carmi jerked the steering wheel to the left. The truck's bumper clipped the motorcycle's rear tire, throwing the rider from his seat.

Carmi, looking back, saw the motorcyclist stagger to his feet. For a moment Carmi thought the man would draw his revolver. But he just stood there, perhaps too dazed, perhaps in too much pain, to do more than watch the truck race out of the city.

That night at Giv'at Hashlosha the crates were removed from the truck and covered with bales of hay. In the morning, a convoy of farm vehicles would take the boxes to a cave near Sharon. The weapons would remain hidden there until the Yishuv had to fight its next battle.

The crack of gunfire echoed each day through the snow-covered hills surrounding Fiuggi. The British gunnery sergeant had drawn a swastika on the targets and he told the soldiers, "Squeeze gently now, lads. Don't pull. Aim for Jerry's heart." Carmi and his men trained with an attentiveness that was nearly devout. He was soon able to put round after round into the center of the swastika. His marksmanship filled him with pride. He looked forward to going up to the front line and shooting at live targets. But Carmi was always aware that it would only be a matter of time before the new skills he was learning would be put to use against another enemy. And that the crates hidden in a cave in Sharon would need to be brought out.

SIX

Peltz, too, was eager to get his men ready for combat. The first snow that blanketed Fiuggi gave him an idea.

After reveille, he ordered his company to dress in their summer khaki uniforms. Marching up the hill to the training area, the men complained. A month earlier they had been parading under the strong desert sun. They were not accustomed to this weather, and the lightweight uniforms made the cold intolerable.

Peltz ignored their murmurings.

As they climbed higher, snow showers started to fall. Soon the flakes were coming down in a thick curtain. The men wanted to turn back. Peltz, however, would not relent. "Come on. Keep going," he badgered. The men continued to complain, but they obeyed.

The next day when they were marching off to the training area, Peltz discovered that one of the soldiers was hiding a sweater beneath his summer khaki shirt. That evening Peltz assembled the entire company. As the troops watched, Peltz made the offender step forward. He accused him of being weak and cowardly. "You are a disgrace to the Brigade," he thundered. Then Peltz sentenced him to a month of guard duty. After that, there were no other attempts to disobey Peltz's dress order.

For the next week, regardless of the weather, Peltz's men trained in their summer uniforms. Peltz genuinely believed it was a way to toughen up his troops, to prepare men from sunny Palestine to chase the enemy up into the frigid Alps if necessary. He wanted Jews to feel that they could stand up to the vaunted Nazi soldiers.

But there was also another, more personal motive for his making such strong demands on his men. He wanted the British to appreciate that although he was a Jew, he was as uncompromising in his standards as any English officer.

Johanan Peltz had never been a Zionist. In his heart Peltz was, like his physician father who had won an Iron Cross for his courage in the Great War, a Polish patriot. But when Jews were expelled from the university in Warsaw, Peltz reluctantly enrolled at the Technion in Haifa. He made the journey to Palestine certain that by the time he earned his degree in engineering, the politics of his homeland would be more rational and he would return.

At the Technion, while his fellow students spoke earnestly about their commitment to the Yishuv, Peltz could find no reason for such ardor in this flat, dusty land. And when some dropped leaden hints about their membership in vital secret organizations, he made it clear he had little interest in their internecine world. He was impatient with those who reduced life into Jewish versus non-Jewish issues. This was, he thought, the worst sort of sentimentality. As a result, he often felt as much a stranger in Palestine as any of the British who had traveled far from their green island to govern a hostile territory populated with squabbling Jews and Arabs.

But before long Peltz, too, came to the attention of the Haganah talent spotters. He had been invited to spend a weekend with classmates at a farm near Carmel. As his astonished friends looked on, he ran full-speed to a bridled horse in the corral, placed his hands firm on the animal's haunches, mounted her, and was off in a gallop. That was *"gonczy stok,"* the messenger's leap, he casually explained over dinner that night.

When the Haganah heard about the young Polish immigrant who could ride like a cavalryman, they summoned Peltz. The Yishuv needed men who could travel on horseback to the remote settlements and protect the inhabitants from the assaults of the Arabs and the indifference of the British. They wanted Peltz to join the Palestinian Mounted Police.

Their argument, an emotional appeal about the responsibilities of Jews to their homeland, had little effect on Peltz. But Peltz was bored with the routine of student life, and in the end he agreed. The Mounted Police would be his last chance to have an adventure before returning to the tedium of his studies. He would serve one year.

Three years later, Peltz was still in the Palestinian Police. He had become well known throughout the territory as "the hero of Sdom," the young police officer who, although wounded, had driven off the Arabs after the ambush of the workers at the Sdom potash plant near the Dead Sea.

Peltz's dispute with the Haganah had started when he had returned from his recuperation at the Hadassah Hospital in Jerusalem to the hills of Sdom. He had come back determined to find the Arabs who had participated in the attack. The local Haganah commander, however, learned of Peltz's search and told him to stop. The official response to the ambush was *havlagah*, self-restraint. If Arabs and Jews were one day to share a state, they would have to stop killing one another.

A *havlagah* future when Jews and Arabs would live together was not Peltz's concern; besides, he would be long gone from this land by then. All he cared about was his duty to the men—his friends—who had been killed. Peltz repeated his vow to the Haganah commanders that he would hunt down the murderers.

Yakov Patt, the Haganah official, announced, "You are no longer to be trusted." Word spread through the close-knit community of Palestine that Johanan Peltz was *poresh*, a man the Yishuv could not count on.

A year later when Peltz volunteered for the British army and was accepted into officers' school, the Haganah sent a message that he was not to go. Peltz disregarded the order. When the Jewish cadets locked Peltz out of the barracks, he smashed through the door with his rifle butt and knocked to the floor the first cadet who dared to block his path. After that he was left alone. He graduated second in his class and was commissioned a second lieutenant in the Palestine Regiment.

Now a company commander in Fiuggi, Peltz still did not care what either the Haganah or the men thought. He was determined to do whatever he believed was necessary to make his troops ready for combat.

And he found an ally at headquarters. After Brigadier Benjamin heard about Peltz's unique dress order, a memo was sent throughout the Brigade: all troops would henceforth be required to train in summer khakis.

When Churchill announced the formation of the Brigade, the War Office decided that it was necessary to have a Jew command the Palestinians. They argued that the troops would only obey a fellow member of their religion. And, no less a concern, it would ensure that the new brigadier would not be instinctively prejudiced against the men.

However, there were no Jewish generals and only three Jewish colonels in the entire British army. After considerable debate, they settled on Ernest Frank Benjamin, an engineering corps colonel and a career officer who at the time was running the Combined Maneuvers Center in Italy.

Benjamin had combat experience, commanding the British troops in Madagascar in their campaign against the Vichy forces. But what won him the appointment was the feeling that, as one fellow officer put it, "he was the right sort of chap." He may have been born in Canada and educated in the red-brick buildings of Clifton College, Bristol, but Benjamin was, down to his monocle, swagger stick, and well-cut uniform, the very model of a British officer.

It was this imprimatur, hopefully stamped deeply onto Benjamin after two decades in the officers' mess, that gave the Sandhurst crowd and the old Etonians in the War Office some confidence in the man, and in their ability to control him. While it had been a political decision to mobilize the Brigade, many in the War Office felt it would be a military blunder, one that would have dangerous repercussions when the fighting in Europe was over, if the Palestinians were taught how to wage war. They counted on Benjamin to understand the tacit limitations implicit in his assignment.

But there was a reality that was not grasped by the high command. It was impossible for a Jew to make a life for himself in the British army without being obsessively reminded of his heritage. Even if he tried to ignore his Jewishness, there were always others eager to point it out. So perhaps it was inevitable that Brigadier Benjamin quickly threw himself into his new command with both the diligence of a career soldier and the intensity, as the old guard in the War Office began sourly to complain, of a man who had scores to settle.

"Training," he insisted in the first official memorandum he distributed throughout the Brigade, "must be of the highest order." He did not care about the political consequences of molding a Jewish fighting force. He was preparing an army to fight the Germans. He wanted to make sure British troops would win. And, his other motivating duty, he was determined to prove that Jews could distinguish themselves on the battlefield.

When reports of the arduous daily training going on in Fiuggi reached the high command, Benjamin had a visitor from the Staff Office. Over lunch the guest made it clear that the feeling at headquarters was that Benjamin was being a little too zealous. "You're British, after all," he was reportedly told. "You're not one of them." Benjamin listened politely. He did not argue. But the next day he summoned the Brigade's senior chaplain, Rabbi Bernard Casper, to his office. He told the chaplain that he wanted to begin daily Hebrew lessons.

SEVEN

The weeks in Fiuggi dragged on. And as the men waited to be called to the front, a cold, muddy hilltop Italian resort was slowly transformed into a Jewish community.

Each morning in the town square the Brigade's blue-and-white flag with the Star of David was raised. Hebrew traffic signs appeared on street corners. With some ceremony, the plaque identifying the Fiuggi Bottling Works was removed and replaced with one in large Hebrew script announcing the headquarters of the Jewish Fighting Brigade. The big Dodge trucks with the Star of David on their mudguards that were constantly rumbling through the cobblestoned streets were painted with playful nicknames; one bore the legend "The Gefilte Fish," as if to boast that a Jewish army travels on a kosher stomach. And with a matter-of-factness that surprised even the troops, Hebrew became the lingua franca of the town. It was not long before many of the locals playing dominos in the cafés would interrupt their games to greet a group of soldiers with a smiling *"Shalom."*

In the frigid December weeks before Hanukkah, some of the men began to construct a six-foot-high menorah on a hill above the town. When Pinchuk was asked if he wanted to help, he brusquely refused. He had no interest in religious symbols, and building a huge menorah seemed a particularly pointless exercise. He dismissed it as child's work.

Besides, he was exhausted at the end of each day. The constant training had worn him down. More than ever he was convinced that Peltz was a tyrant. And worse, he had begun to realize that the brigadier shared the same vindictive temperament. When Pinchuk's day was done, he just wanted to sit on his cot and read. He was dutifully making his way through Shakespeare's tragedies, and this was his only comfort.

Seven years earlier Pinchuk had fled from Reflovka when a policeman had confided to his uncle that Arie was going to be arrested in the latest roundup of "Communist sympathizers." A long jail sentence was the certain punishment. With five pounds sterling hidden in his coat, Pinchuk, a skinny, bookish seventeen-year-old, left the next day. His parents and his sister could not stop crying as the train pulled away from the village station. But Pinchuk, making a daring escape to Palestine, was elated.

He arrived in Haifa in October 1937. The heat was unbearable, and the land was foreign. At his first meal he was given a plate of bitter green olives and he popped a handful in his mouth thinking they were cherries. But he also found opportunities in his new home. He enrolled in the university in Jerusalem and immersed himself in his studies, excited to study philosophy in a classroom rather than in afternoon chats in Reflovka's barbershop.

And he was certain that his homesickness would be short-lived. His parents had promised that it would not be long before the family would be reunited with him in Palestine. A year, perhaps two at the most, and the money necessary for the visas would be saved.

Then the war broke out, and the Soviet army marched into the Ukraine. Yet he remained hopeful. His father still sent news-filled letters reassuring him that despite the distant war, life in their little village continued in its complacent way. But in June 1941, the Nazis invaded the Soviet Union and the Red Army retreated from the border territories. And the letters stopped.

At first Arie tried not to articulate all his anxieties. He felt that if he were to put his fears into words, he would increase their awful likelihood. But as the rumors of what was happening in the German-controlled Ukraine began to spread to Palestine, Pinchuk came to a wrenching realization. He was a deserter. He could not forgive himself for leaving his family behind. His sole hope for redemption was to get to Europe, to get back to Reflovka and protect his family. The only way he could think to do this was to join the British army. An unlikely soldier, Pinchuk volunteered and then blustered his way into officers' school.

Now Pinchuk was in Europe, a lieutenant in the Third Battalion of the Brigade. The first part of his bold plan had succeeded. But he was trapped in Italy for the winter. All he could do was wait, and hope that it was not too late.

On the first night of Hanukkah, when the candle was lit in the oversized menorah the men had erected, Pinchuk was making his way through the town square. He was returning to his room from dinner and looking forward to settling into his cot with his volume of Shakespeare. It was simply a matter of chance that he happened to observe the ceremony, but he found he was caught off balance by the intensity of his response. He stared up at the ancient Jewish symbol shining brightly in the night on the Italian hilltop, and a sudden link was forged between his private and public missions. To him, it seemed as if a beacon reaching out across war-ravaged Europe had been illuminated, a signal proclaiming, *Jewish soldiers have arrived. Help is on its way.*

EIGHT

The winter lull in the fighting gave the men the chance to explore a bit of Italy. When Peltz stopped Pinchuk one morning in December and asked if he wanted to join Carmi and himself when they took leave in Rome, Pinchuk was surprised. Ever since they had begun training, Peltz had ignored him except to shout orders. This was fine with Pinchuk; any further conversation and all the angry, insubordinate thoughts he had might have spilled out.

But once the invitation was offered, Pinchuk decided it would be petty to refuse. The opportunity to see Rome, and to spend time in Carmi's company, was too appealing. He petitioned the battalion commander for a thirty-six-hour leave, and it was granted.

The three men hitched a ride in a truck heading north for provisions. The road was muddy, deeply rutted in spots, and it was a long, jerking trip. The three soldiers, though, made themselves comfortable in the back. As they began to talk, Pinchuk felt as if they were on the *Stafford* again in heaving seas. Encouraged by that memory, he found it hard to hold a grudge.

In time, Pinchuk shared a story that was making its whispered way around the battalion. One of the British officers, a Maj. Forge, had refused to wear the Brigade's cap with its gold Star of David. Instead, he showed up for parade with the cap of his former Scottish regiment on his head. "Damned if I'll wear this bloody Jewish thing," he had complained.

The next day the major's cap was stolen from his room. It was returned two days later—burned, the charred remnants scattered on the major's bed. The MPs were looking for the culprit, but so far they had only been able to narrow the suspects down to several thousand Jewish soldiers.

"Hope they never find the man who did it," said Pinchuk. "Serves the anti-Semite right."

"What kind of man steals a hat?" Peltz argued. "Ridiculous. You want to teach an anti-Semite a lesson, you teach him so he won't forget."

And as the truck continued on to Rome, Peltz told them what had happened when he had been serving with the Palestinian Police in the rural village of Hartuv.

"There was a Welshman stationed there by the name of Goult. A rough and unpleasant type. Full of British colonial superiority. One day I was sitting on the front verandah of the police station reading a book when Goult walked up the stairs. The station was really an old stone house that had previously belonged to a wealthy man, Levy, who had gone to live in Jerusalem. Well, that day there was a slight breeze and the wind pushed aside some vines covering a Magen David cut in the stone above the entrance. Goult stopped, looked up at the Magen David, then at me, and said: 'I didn't know that I am living in a fucking Jewish house with this fucking Jewish star over the front door. I'm going to chip it off.'

"So I got up from my chair, folded it up, and without a word I smashed the chair over Goult's head. He collapsed to his knees but I kept kicking him in the ribs. He rolled down the steps, and while he was lying on the ground I stood over him and shouted, 'If you damage this Jewish sign in any way, I am going to break you into pieces.'

"I wound up getting an official reprimand and fined five days' pay, but it was worth it. That's the way you handle anti-Semites."

For the first time, Pinchuk felt a rush of respect for his strong-willed superior.

"A few days after the incident," Peltz continued, "one of the British policemen in the station came up to me. He was Harvey-hyphen-Gordon, another of those better class of Britishers who for one reason or another found themselves serving in Palestine.

" 'I am not saying you were wrong, Peltz,' he told me. 'But the way you went after Goult was entirely ungentlemanly. You should have challenged him to a fair fight.'

" 'Does that mean you are challenging me to a fair fight, Mr. Harvey-Gordon?' I asked.

" 'Yes,' he said. 'I do. You need to be taught a lesson.'

"So the fight was held in the backyard in an improvised ring. We were both about the same size and weight. Harvey-Gordon was some years older.

"But for me the fight was a disaster. I never landed one single punch. It seemed Harvey-Gordon had been a member of a boxing team in some English college. He never once hit me in the face, but he could have. He just kept landing sharp, painful blows on my chest and ribs. After two rounds, they stopped the fight. I lay on my bed for a whole day aching with pain. My ribs were so sore I could not take a deep breath for weeks.

"The next day Harvey-Gordon came to visit me. He told me, 'No hard feelings, John, but you had to be taught a lesson. I don't blame you for the Goult incident. But you went over the line using a chair and kicking him when he was down. As an Englishman I had to put the record straight.'

"Well, when I thought about it, I began to realize Harvey-Gordon had a point. Strangely enough, Harvey-Gordon and I became quite friendly after that. We even got together several times after I left Hartuv."

Now Carmi spoke up. "Know what I would have done with your Mr. Harvey-hyphen-Gordon? I would have smashed a chair over his head, too. And kicked him silly when he was down. Know what's the trouble with you, Johanan? You like the English too much. You haven't learned that Jews don't have any friends."

The argument continued as the truck headed toward Rome. But

Pinchuk did not join in. He was too deep in his own thoughts. He was beginning to have a sense of how different he was from these two men. His own weakness embarrassed him. And he wondered if he could even summon up the courage to steal some bastard's cap.

In Rome, though, Pinchuk's confidence returned. Walking through the recently liberated streets in his British officer's uniform, it was easy to put aside all his concerns. He had not fired a shot, but he felt like a soldier in a conquering army.

It was a quick, fast-paced visit. There were conversations in broken, rudimentary Italian with smiling dark-haired girls whose names he never knew. He ate bowls of spaghetti covered with an oily red sauce, and greedily soaked up the remains with pieces of crusty bread. He shared bottle after bottle of ruby-colored wine. But there was one moment in his tour of the city that, after all the other memories were lost in a happy jumble, remained clear and affecting.

The three friends were walking down a strange street, sightseeing, when a priest dressed in his black cassock approached from the opposite direction. He nodded a silent hello, but then he noticed the Jewish stars on their shoulder patches. He became excited. You're Jewish? he asked. Soldiers?

"Yes," Peltz answered, on guard.

Come, said the priest, pointing toward an ancient church across the way. Please come.

"He wants to convert us," Pinchuk warned.

"Let him try," Carmi said.

The priest led them into the church and down a flight of stone steps into the dark basement.

This church, he explained in slow but precise English, is called San Pietri in Vinoculi—St. Peter in chains. Does that mean anything to you?

They shrugged. The priest seemed surprised by their ignorance. But he did not offer to explain.

The priest led them to a corner of the basement, and pointed at a pile of burnished metal chains. Please, the priest urged. Pick them up.

The chains made a jangling noise as Pinchuk lifted them. He had not expected them to be so heavy, and the links felt cold, almost damp in his hand.

"How old are they?" he asked. He was being polite.

The priest explained. When the Romans conquered Judea in A.D. 70, the Jewish prisoners of war were brought back to Rome the following year in chains to be sold as slaves. These were the chains.

The three soldiers thanked the priest for showing them the artifacts, and then hurried back to the street, and to the wine, and to the dark-haired girls.

But Pinchuk never forgot what he had seen. His ancestors had been bound by those chains. And he wanted to believe that with the arrival in Europe of the Brigade, a Jewish army, the pattern of victimization that had plagued his people throughout history was about to be broken.

On the ride back to Fiuggi, as Peltz and Carmi slept, Pinchuk found himself still deeply moved by the experience in the church. It was as if he could still feel the weight of the cool heavy chains in his hand.

After a while his thoughts turned to the Shakespeare play he had been reading before they had left. It was *Henry V*, one of his favorites. He had first studied it as a boy, and he had been swept along by scenes of saber-rattling adventure. But it was not until this moment that he fully understood what he had read. The words he had memorized as a child resonated in his mind:

> We few, we happy few, we band of brothers;
> For he to-day that sheds his blood with me
> Shall be my brother; be he ne'er so vile
> This day shall gentle his condition:
> And gentlemen in England, now a-bed,

Shall think themselves accurs'd they were not
here,
And hold their manhoods cheap whiles any speaks
That fought with us upon Saint Crispin's day.

As he recited these words in his mind, Pinchuk found himself moved by his companionship with his own band of brothers. And no less deeply, by the honor of going off alongside his fellow Jews to fight the Nazis.

But back in camp, alone again in his own bed, Pinchuk was tormented by his fears about his family. When confronted with this reality, honor became an illusion. And his hope collapsed. He found it hard to believe that all of them—Mama, Papa, Leah—were not already dead. He suspected that although he had come to rescue them, he was too late. He had failed.

NINE

Leah Pinchuk was afraid to sleep in the dark. She was thirteen years old but she insisted that her mother light the candle next to her bed. When the Germans took control of Reflovka in the summer of 1941, her biggest fear was that they would order all candles to be extinguished at night.

But nothing seemed to change drastically. Her father, Meir, and her mother, Reitze, both continued to work in a government office the Russians had established to distribute food. Leah went to school. And at night a candle glowed reassuringly by her bed.

After the High Holidays in the fall of 1941, however, the German district commander in Sernie issued a decree: All Jews in the area were to be relocated into a ghetto. It was to be established in the center of the village of Reflovka.

The convoy of horse-drawn peasant carts that brought the families from Zlotchk, from Olizerka, from Bielskovolia, stretched far down the dirt road leading into the village. Over 2,500 people, according to the count made by the methodical German officials, were crammed into the circle of houses surrounding the synagogue on Pillsodesky Street. Fourteen people, a family that spanned three generations, were ordered to move in with the Pinchuks. It was a three room house.

* * *

The ghetto was open. There was no fence. But it was impossible to leave without a permit. The Ukrainian police stood guard.

The Jews were required to wear an identifying yellow star, precisely nine centimeters in diameter, on their topcoats. Another yellow star was to be displayed on clothing covering the chest.

Leah cried when her mother sewed the yellow star on her black school dress. The uniform had made her feel very mature, as if she were one of the older girls who already had boyfriends. Now it was ugly, and she felt diminished, too. She was thirteen years old and wanted very much to be pretty.

Soon Leah's school days were interrupted. All children in the ghetto enrolled in the third grade or higher were required to serve in labor details. Leah was ordered to join a group of women who sewed wool hats, gloves, and socks for the German army. But work on the wooden replacement for the Satir River bridge that the retreating Russian troops had destroyed had fallen behind schedule, and the Reflovka girls were assigned to supplement the existing crews.

Leah had been working on the bridge for several weeks when, on the day before she was to go back out, the crew from Reflovka did not return. There were sixty men and boys on the bridge detail that day, and the Ukrainian guards had decided to execute them. No reason was ever given.

The Judenreit, the ghetto's council of officials, warily explained to the local German commander that these executions would only delay the completion of the bridge. He agreed. The work crews would no longer be harassed by the guards, he assured them.

Nevertheless, the next morning Leah's mother attempted to take her daughter's place in the detail. The Ukrainian guard would not let her. Reitze begged, cried, offered money. The guard insisted the girl had to leave with the work party.

Reitze sobbed as her daughter walked off with the others. Leah was crying, too. An older girl at last had to lead her away.

When Leah returned that evening her mother was waiting on the street corner where the ghetto began. The two hugged, and cried, and then began to laugh. That night at dinner there were strawberries and cream, Leah's favorite treat, for dessert.

Each day was woven together with its own pattern of hardships. The only way to get through it was to convince yourself it would not always be like this. The war would end one day, and all this, thanks to the grace of the Almighty, blessed be His name, will have passed.

But at the beginning of the last week in August, 1942, a rumor spread through the ghetto: Outside the village of Sachovola, just three kilometers from Reflovka, crews were excavating two large pits. The work was nearly completed.

At first people offered many reasons why the Germans would want to dig two large pits. Then on Wednesday morning, the group leaving to work on the bridge was ordered back to their homes. Later that day the Judenreit was informed that a census would be conducted. On Saturday morning everyone was to assemble in the town square. No one was allowed to leave the ghetto.

There were no longer any doubts about why the two deep pits had been dug.

Leah's uncle had a plan. He had been deliberately cultivating a friendship with one of the Ukrainian policemen assigned to his work force. For months he had been hinting to the policeman that, despite the many *contributzias* that had been demanded, he had something special hidden away. The time had come to disclose the details of this secret to his "friend."

I have a box filled with gold, he told the policeman. It's buried beneath a tree. No one will ever find it. But if you will help my niece escape, I will tell you where it is.

The policeman agreed. He would be waiting at midnight on a street adjacent to the ghetto. Bring the girl, and he would escort her to the forest. After that, she would be on her own.

When the Pinchuks were told, they rejoiced. But Leah could not be convinced. She had heard a story about a girl who had escaped to the forest only to be raped and murdered by her Ukrainian savior.

Both her mother and her uncle pleaded with her: "It's your only chance."

"Your fate is my fate," Leah told her mother. "Whatever happens to you will happen to me."

But Leah's father also had a plan.

Late Friday night, after his wife and daughter had gone to sleep, he sneaked into the yard behind his brother-in-law's house. Next to a row of trashcans, in a dim corner of the plot, was a large box filled with animal dung.

He pushed the foul-smelling box aside, took his brother-in-law's shovel, and began to dig. When there was a hole wide and deep enough for two people, he stopped. Then he moved the box with the animal dung over it.

At first light he led Reitze and Leah into the yard. His wife clasped one of her husband's hands. His daughter held the other.

"Listen to me," Meir ordered. "Do what I tell you. No matter what happens, you are not to move from this hiding place until it is dark. Until after midnight. Do you understand?"

Both his wife and his daughter were trying not to cry. "No matter what you hear, you are not to move. Do you understand?"

They managed to say yes. Meir hugged his wife and his daughter.

"Remember," he told his wife, "when this is over, you will go to Arie in Palestine. He will take care of you."

He turned to his daughter. "What's your brother's address?"

"Sixteen Betzalel Street," Leah answered.

"My smart girl." He gave Leah a last kiss and helped them into the hole. When they were inside, he moved the box with the dung on top. He made sure there was a small space, a sliver, between the wooden box and an edge of the hole; it would provide them with air.

Through this narrow aperture, Leah watched her father position the garbage cans around the box. She saw him look back at their hiding place, then hurry away without another word.

When her father had moved the box over her head, Leah thought this was what it must be like when the coffin lid is dropped into place.

But she did not dare to climb out. Huddled in the dark, foul-smelling hole next to her mother, Leah listened:

Mrs. Chodler urging people to come out of their hiding places. You can't hide, the old lady shrilly repeated.

A rattling on the garbage cans. Someone was trying to climb inside. She recognized her aunt's panicked voice. Then gunshots. And the sound of her aunt's body being dragged away.

The orders being given to the people assembled in the square. Move out, the amplified voice explained. You are being relocated.

The Ukrainians running into their houses. There was breaking glass, excitement, and laughter.

Then, finally, a vast, empty quiet.

It was all there was until Leah's mother whispered, "Come. We should go."

Leah climbed out of the hole. It was night. And then she was running as fast as she could, following her mother toward the thick dark forest.

TEN

CONFIDENTIAL

JEWISH INF BED GP INT SUMMARY

Dated 1 Mar 45.

1. ENEMY SITUATION

The present Russian offensive naturally gives rise to speculation as to the future of the German Armies in Italy. The overall strategy of the High Command does not, however, appear to have changed—that is, to hold as much as possible of the German defense lines. There is an extensive program in existence which would indicate a defense of no ordinary tenacity of the resources of Northern Italy. It looks indeed as though FIELD MARSHAL KESSELRING will withdraw nowhere unless he is pushed, and pushed hard.

Peltz finished reading the intelligence report, and underlined the final sentence. He used a ruler to make certain the line would be straight; that was how he had been taught at the Technion. Then he went looking for Pinchuk.

He found his friend playing bridge with a group of officers in the dimly lit hotel salon.

"I want to show you something. Put the cards down, Arie," he said.

"What's so important?"

"Just finish up."

Since their return from Rome, Pinchuk had spent enough time with Peltz to be unsurprised by such peremptory behavior. He had never encountered anyone filled with so much self-confidence, or as oblivious to the feelings of others. Peltz's manner was so different from his own tentative, inhibited way that it made him fall into long periods of self-analysis. And the more Pinchuk made comparisons, the more he began to find something instructive, even liberating, in Peltz's example. To his astonishment, he found that he wanted to be able to stride through the world like Peltz.

So Pinchuk, set on making a point, played out the rubber. When he was done, he joined Peltz, who had been pacing impatiently on the other side of the room.

Peltz handed him the page. "Read it."

Pinchuk read in silence, then looked up questioningly at Peltz.

"You read that?" Peltz asked. " 'Pushed, and pushed hard.' "

Pinchuk heard his friend's excitement, but he still did not understand.

"I was at Command this evening. That's where I got this brief. Know what they were talking about? *Trucks.*"

Suddenly Pinchuk too was excited.

"Trucks, Arie. I'm telling you, it's time. The Brigade's going to be moving out soon. 'Pushed, and pushed hard,' Arie. We're going to get the chance to fight after all."

All through the winter, the front had been quiet. South of the graceful medieval city of Bologna with its Gothic brick churches, red tiled roofs, and porticoed squares, the two opposing armies—nearly a half-million men—were positioned on either side of the winter line.

It was a sinuous, often meandering front, determined more by the natural authority of geography than by the cunning of generals. It started high in the snow-capped Apennine hills; hugged the banks of a series of icy, gray-blue Romagna rivers; rose up into a spike of hilltop ridges only to dip with dramatic suddenness as it crawled across large, flat frozen plains; then followed the long, curving southern shore of wind-swept Lake Comacchio; and finally cut haphazardly through the bleak marshes that led to the Adriatic. To the north, were the twenty-three divisions of the German Army Group C. South, in nearly equal strength, were the Allies, the Fifth and Eighth armies.

Field Marshal Albert Kesselring, commander-in-chief of the German army in the Mediterranean and southeastern Europe, had returned from Berlin to take command in January. He had no illusions about the inevitable course of the war. Or the tenacity of the force poised against him.

The Allies had already broken through his impregnable Gustav Line and, two days before D-Day, had marched into Rome. Methodically they had pushed his army farther and farther back, until the Wehrmacht was fleeing like fugitives into northern Italy. Then winter came, and the fighting stopped.

Kesserling knew that as soon as the snow began to melt and the rain subsided, the Allies would once again charge forward. When the spring offensive began, the enemy would persist until either he surrendered, or his army was destroyed.

But Kesselring had his orders. Hitler, still intent on fighting on at any cost, had sent him back to Italy with the unequivocal command to "defend every inch of the north Italian areas." With a soldier's fatalism, the field marshal returned to his post determined to obey.

Yet as he spent the long, quiet winter analyzing the situation, he became convinced that he could put up an effective fight. He still did not pretend to himself, or to his generals, that he could reverse the ultimate course of the war. But he did believe that he could postpone the defeat of the Wehrmacht in Italy. Perhaps he could even manage to hold on until another winter brought the fighting to a halt. At the very least, he would make the Allies win their final victories.

This newfound sense that he could achieve a proud and honorable stalemate was encouraged by two elements. First, he had resources. Army Group C was up to strength and, after the winter peace, well rested. On the line, were the First and Fourth Para Divisions, combat units toughened by fighting in both Russia and central Italy. The infantry divisions, while filled out with Austrian and Italian troops whose commitment to the Reich could prove tenuous, had a loyal core of German veterans. Moreover these divisions were fully staffed, not the thin, depleted battle groups that were unrealistically still identified as "divisions" on the other German fronts. In addition, he had hundreds of serviceable tanks—another unexpected resource this late in the grueling war—and two well-trained panzer grenadier divisions to operate them. Fuel, like everywhere else in the German lines, was a problem. But he had plenty of food, ammunition, and men. Army Group C, he was convinced, was prepared to fight back.

And, his other reason for hope, he had a plan. It was called Autumn Mist, and it was a strategy of lulling deception. His men would need to endure for months; the more the Wehrmacht suffered, the more unsuspecting the enemy would grow. But Kesserling promised his generals that by the first cool days of autumn, the "mist" would evaporate; and with clear and stunning power the plan's lethal guile would be revealed.

The initial stages of Autumn Mist, however, would be a series of strategic withdrawals by the Germans. Kesselring would hold on to the river positions south of Bologna for as long as possible. Then, undaunted, he would pull back to a new, well-fortified line that curved north of Bologna along the Reno River. From the high ground, his Spandaus and tanks could fire with brutal accuracy. But when pushed, he would abandon this position and retreat to a third defensive line along the broad banks of the Po River. The Allies would have to pay a bloody price to drive him back from the Po, and their initial attempts to cross the wide river would be nearly catastrophic. Still, Kesselring had no doubt they would press on to accomplish this feat. They would bravely charge after the retreating Wehrmacht. And into the death trap he had cunningly set.

Throughout the winter, under the goading direction of General Buelowius, the Reich's inspector of land fortifications, five thousand German construction specialists assisted by thousands of Italian laborers worked around the clock to finish the position where Army Group C would make its last stand. The fortified line ran for miles, starting east of Lake Garda and then moving across the steep Alpine foothills and down over the plains north of Venice. Its high, reinforced concrete walls were laced with curtains of razor-sharp barbed wire. The approaches were littered with one thousand-pound mines. It was a twisting warren of machine gun pillboxes and mobile cannon emplacements. Huge, long-barreled artillery guns commandeered from the Ligurian coast were mounted at its highest points; their firepower was tremendous. The Berlin Command, in a burst of Germanic martial prose, christened the stronghold "The Forward Alpine Defenses of the National Redoubt."

For Kesselring, it was the great gamble of his career. If he won, his army would survive to rest for another winter. And to fight for another spring.

But in every way, it was a nasty, desperate wager. He was using his own men as bait as surely as if they were goats tethered to a stake. He hoped to find consolation in the revenge he would take on an enemy so filled with blood lust that, despite the mounting carnage, despite any instincts for self-protection, would chase on and on. Straight into Hell.

General Sir Harold Alexander, commander-in-chief of the Allied army group in Italy, was by breeding an aristocrat and by training a strategist. But there was little that was genteel or inventive in his design for the spring offensive. His plan was rough and straightforward. He would drive the enemy back against the banks of the Po and, as he succinctly promised, "annihilate him there."

With one sustained, concentrated thrust the Allies would make the Forward Alpine Defenses of the National Redoubt a preposterous irrelevancy. The Nazis would never have the opportunity to brace down in their harrowing fortifications for a grand, Wagnerian last stand. It would be too late. They would already have been defeated.

But if this strategy were to succeed, the Allies had to cross the network of Reno River defenses—each one a deadly obstacle—with speed and precision. They would need to drive the Wehrmacht back from these front lines so swiftly and so forcefully that, broken and demoralized, its troops would be unable to regroup to cross the Po.

Instead, the battered German army would be caught in the flat plain between the two rivers. Then in one galvanized explosion of Allied might, the campaign would end. This was the perfect killing ground.

It was, at least on paper, an attractively simple plan, and Alexander spent much of the winter working out the tactical details. He decided that the Fifth Army, employing a classic outflanking maneuver, would attack west of the Reno, hooking toward the enemy's rear defenses. Simultaneously, the Eighth Army would charge the heart of the Romagna river defenses.

But no sooner had Alexander articulated this broad strategy to his generals than he began to realize how much its success hinged on the Eighth Army's swift penetration of the enemy lines. And how difficult these river crossing would be.

In the spring, many of the rivers would be running high. A further complication was that these deep swirling waters would need to be forded under fire from an enemy who occupied the high ground, and who had had an entire winter to reinforce his position.

Yet if the Eighth Army was repulsed, the Germans would escape across the Po, and the outmanned Fifth Army would be dangerously exposed.

With so much at stake, several alternative plans were considered. A massive parachute drop of men into the plains beyond the river network was proposed. For a while it seemed the perfect solution.

But when the weather cleared, reconnaissance photographs revealed strong German air defenses in the area. The Allied air forces insisted their bombers could destroy these guns; however, after much agonizing, Alexander vetoed the idea. It was filled with too many risks, too many chances for failure. If the planes carrying the parachutists were shot down, the entire operation would be doomed.

Another proposal involved an attack force that would forge across the swirling waters in an armada of Fantails. These tracked amphibious landing vehicles had been used with great success during D-Day. Except as soon as Alexander began to give this serious consideration he learned there were only four hundred Fantails in Italy, and the likelihood of acquiring additional vehicles before the offensive began was at best problematic.

In the end, Alexander decided on a variation of his original plan. Instead of the Eighth Army making a unified, massed assault through the Argenta Gap toward the Reno River, they would make a series of preliminary thrusts. This would serve two tactical purposes.

First, the Germans might be deceived into believing that these were the main attacks. If they fell for the trick, Kesselring would hurriedly begin moving defenders into these positions, and away from the actual primary attack site.

Second, if the river crossings were successful, the troops would be able to sweep east, behind the Germans, and support the main Eighth Army force from the rear.

Of course, the troops making these preliminary crossings would be in enormous danger. If the deception succeeded, the panicked Germans would deploy an overwhelming force against the much smaller units fording the rivers. Nevertheless, Alexander decided he had no choice.

The key crossing in this preliminary attack would be over the Senio River. It was a narrow, muddy river, often quite shallow even in spring, with a flat marsh on one shore and steep banks rising toward the hills on the opposite side. The Germans had spent the winter fortifying the high, northern banks with machine gun pillboxes, artillery, and mortars. The approach was densely mined. However, if

Alexander's grand strategy was to succeed, and the Germans pushed back toward the Po, it was vital that the Senio River defenses be decisively overrun.

It was decided that the Jewish Brigade would be given the assignment of crossing the Senio River.

Two days after Peltz had shared his news with Pinchuk, the Brigade moved out from Fiuggi. They traveled north in a long convoy of trucks. The Italian countryside was dormant, colored in dull, brownish shades. The roads were muddy trails. But the men were excited. They had been waiting for so long, and at last it was happening.

Carmi sat in the back of one of the big Dodges with the men from his platoon. The soldiers were singing and he was reminded of the day they had left Alexandria. As the road climbed closer to the front, the convoy passed a vehicle carrying a British unit going in the opposite direction.

Carmi noticed how curiously these soldiers looked at his men. They did not understand how troops heading off to fight could be filled with a sense of joy. Carmi did not blame them. How could they understand what it was like to be a Jew anticipating the chance to confront his blood enemy?

Throughout the first week in March, the Brigade took up positions on the battle line outside Ravenna. The offensive would, they hoped, begin soon.

PART III

THE FRONT

Spring 1945

ELEVEN

The country was very flat, crossed almost laterally by Canals and Fosses, four of which lay between the Bde GP and the River SENIO.

In addition many small streams and irrigation canals joined the major waterways at irregular intervals across the entire Bde front. The importance of these canals lay not so much in their value as obstacles but in their high banks. . . .

The JEWISH INF BDE GP was to carry out active patrolling and improve its position as opportunity arose, but at this time was not intended to mount any set-piece attack.

—Official War Diary, Jewish Infantry Brigade Group

Capt. Johanan Peltz did not dare to lift his head. His chin was pressed into the soft, moist dirt. He lay there listening. The noise from the barrage of German shells was nearly constant. They were being launched with a regular, hammering rhythm. He could track the roaring groan of their flight. And he would wait for them to land. On impact the ground would tremble, and the vibrations would course through his outstretched body as if they were an electric current. But this sensation was also reassuring: the shell had missed.

For nearly two hours—an eternity when spent anticipating the moment that will end your life—Peltz and his men had been pinned down by a barrage of artillery, mortar, and machine gun fire. "We need smoke. Give us some bloody cover," he had screamed into the radio when the Spandaus first began firing. But there had been no answer, and his radio operator was still trying to get battalion headquarters to respond. It was Peltz's third day on patrol along the banks of the narrow canals adjacent to the Senio River.

The patrol, an inspection tour of the front line Peltz's B Company was taking over, had gone smoothly enough at first. In this sector near the small farming town of Alfonsine, the Germans were positioned on the northern sides of the canals, the British to the south. A series of wide, grassy, flat-topped dikes ran intermittently across the waterways; these were no-man's land. All was quiet. Both sides seemed to be waiting for the offensive to begin.

After consulting his map, Peltz had moved toward the Fosso Vetro, a long irrigation ditch filled with a film of brackish water. His men were spread out in a line along the ridge. And without warning the shelling had begun.

"Take cover," he had shouted. Before he hit the ground, he saw the gray, ominous puffs of artillery coming from a hilltop on the Allied side of the canal. In those first moments he feared that he was being fired upon by his own men, and that his life would end because of a mistake. But by the time he had collected himself to call in for smoke, he had checked his map and worked out what had happened.

Sometime last night, or perhaps even the night before that, the Germans had secretly established themselves in the high ground on the Allied side of the dike. The map identified their new position as La Giorgetta and, Peltz had to concede, it was a shrewd acquisition. From this vantage point the concealed German guns, the dreaded eighty-eight he suspected, could control the entire eastern flank of the sector.

He could not move forward. Even a retreat would be costly unless it was executed under a cloud of smoke thick enough to disorient the German gunners. All Peltz could do was wait. For the first time after years of a disciplined stoicism, he found himself starting to pray.

It was a long, agonizing wait. Finally, the battalion responded. There was smoke, and he was able to haul himself to his knees and crawl back toward a protected position. Once he was out of danger, Peltz settled into a deep, fuming rage. How could Command's intelligence have been so faulty? How could they have allowed him to lead his men without warning into such jeopardy?

But the largest share of his anger and disappointment was aimed at himself. He had been afraid. When the German guns pounded the ground around him, he had been utterly helpless. He had fought and killed men in Palestine, but those battles had been, Peltz was convinced, heroic. In his romantic mind, they might have been duels. But this was another kind of combat: impersonal, mechanized, and brutal. And he began to realize that the war against the Nazis would be different from any fighting he had ever known.

The first time Peltz had ever killed a man was on the night of January 3, 1938. It was a night when he shot many men, and he became "the hero of Sdom."

He had been a twenty-one-year-old police sergeant assigned to accompany an emergency repair crew to the pumping station at Ein Arus, a spring near the southern tip of the Dead Sea. The flow of water to the region's potash plant had mysteriously stopped earlier in the day. It was probably just a blockage in the intake line, but until it was corrected the plant had to be shut down.

A convoy of three vehicles left after midnight. Their headlights illuminated a narrow dirt road. A wall of pale, humpback hills rose in the distance to one side. On the other was a fragrant tangle of dry brush. They continued on this road for a few hours until they were forced to stop. A barricade of large stones blocked their way.

For the first time Peltz seriously suspected the pumping station had been sabotaged. But they had come too far to turn back. And the flow of water to the plant needed to be restored. Peltz announced that he would take two of his men and hike through the brush to Ein Arus. The workers were to wait in their trucks until he returned from his inspection.

Peltz and the two officers proceeded slowly through the thick, bristly undergrowth. The first light of the new day was making it less difficult to see what was in his path, but Peltz found himself thinking that also meant he was an easier target.

Alert and tense, the men continued to move ahead. Then they heard the gunshots. Rapid fire was coming from the direction of the trucks.

"Back," Peltz shouted. Without further command, the three men ran full-speed through the brush, rushing toward the gunfire and their friends.

All at once, as if a mirage, Peltz saw two men standing in a clearing about twenty yards away. His first instinct was to go for his rifle, but then he recognized the Bedouins. They were Fayiz and Salammeh, two of the potash company's watchmen.

"Ya Salammeh!" he called out. "What are you doing here?"

Salammeh did not answer. He raised his Mauser rifle to his shoulder and took aim.

Peltz dropped to one knee. There was no time to get his rifle off his shoulder, and he drew his heavy Colt from his belt as the Bedouin's shot went over his head. Then Peltz fired. And fired again. Salammeh fell to the ground.

The other Bedouin turned and ran. Peltz did not chase after him. It was more important to get back to the others. When he passed Salammeh, he broke stride only long enough to see that both his shots had scored. There were two red circles spreading across his chest. The man was already dead.

By the time Peltz reached the trucks, the Arab attackers had managed to isolate the men from the plant into separate groups. Two of the workers had taken cover under a truck, which was surrounded by at least fifty Arabs. The rest were pinned against a hill, shooting from behind a shield of rocks.

There was no hope of getting to the truck. Firing all the way, Peltz and his men worked their way to the rocks where the others were huddled. But the confined position gave them no advantage. If the Arabs charged, Peltz realized, they would be trapped against the hill. It would be a massacre.

"Up the hill! Quick!" he ordered.

It was a nasty climb. Bullets ricocheted off the rocks, flying in all directions. Peltz hoped to outdistance the Arab rifles, but there seemed to be no escaping their range. Finally he made it to the top.

Now he fired down at the ambushers. He shot with his rifle at his hip, not even trying to take aim. He just wanted to scare the Arabs, to get them to retreat. He pulled the trigger as fast as he could. Suddenly, he felt a sting just below his left knee. But he kept firing.

The Arabs grouped at the base of the hill began to pull back. But as they did, another wave rushed the truck where the two men were hidden. Dozens of Arabs climbed on the truck, banging on it with their rifles.

Peltz and the others on the hilltop held their fire. They could not risk hitting their comrades. But when Peltz saw a few of the Bedouins slithering under the truck, their *shubrya* knives clenched in their hands, he knew he had no choice. "Fire at the Bedus. They are killing our people," he ordered.

So they opened fire. And they kept shooting. Peltz was certain he had killed at least three men. But despite the barrage of bullets, the Arabs swarmed over the truck; until with one final shrill yell of triumph, the Bedouins ran off into the distant hills.

The next day Peltz's leg had swollen dramatically, and he was unable to walk. He was sent on a special fast boat to a doctor on the north end of the Dead Sea. Accompanying him was the one survivor of the attack on the truck. He died of his wounds before the boat docked.

When a paper in Tel Aviv ran a story about the attack on the potash plant workers, Peltz was called "the hero of Sdom." He sent the clipping to his grandfather in Zabiec with a note. "See," he wrote, "I *am* your grandson."

But in Italy he was fighting another war, against another enemy. On the way back to his company's camp in the high ground above the river, Peltz could not get over the shameful feeling that he had been run off. He had never retreated from the Arabs. Yet in his first encounter with the Germans, he had literally crawled away. He had not even fired back. He could not allow the Nazis—it was impossible for him even to say "Nazi" without his bile rising—to get away with such an easy, demeaning victory. And he needed to prove to himself that he was not afraid.

By the time he walked into his tent, his mind was already set. He would retake La Giorgetta.

"You're crazy," Pinchuk promptly told Peltz when he heard his friend's plan at dinner. "This isn't your own personal war against the Nazis."

At Company headquarters, his superiors were also discouraging. One British officer sipped from a flask of gin as he told Peltz, "Not our bailiwick, old boy. Strategy is for the deep thinkers at headquarters. We're cannon fodder. We stay put 'til they tell us to march."

So Peltz went up the chain of command to Col. Geoffrey Gofton-Salmond, the battalion commander. By appealing to a higher officer, Peltz knew he was treading very close to insubordination. But Peltz was certain he was right. In this inflated mood, he could not be deterred.

"Goofy," as Col. Gofton-Salmond was called when he was out of earshot, was a veteran soldier. Before joining the Brigade in Fiuggi, he had commanded a British parachute battalion through some rough fighting in Salerno. Peltz was convinced a man with that kind of experience would not turn him down.

When Peltz proposed to attack the hilltop German fortress, the colonel did not cut him off. He listened silently. Peltz had no clue as to what he was thinking.

The colonel finally asked one question: "What did the major say?"

Peltz told him.

The colonel considered this. At last he spoke. "Don't worry, my boy. I'm not going to throw you out. What I think I will do is send for some tea and biscuits. Let's have a cup and talk this through. Then we'll see."

They talked for over an hour. By the time the pot of tea was drained, they had reached an agreement.

"Danger here," Gofton-Salmond patiently explained, "is going in half-cocked." He ordered Peltz to devise a specific plan of attack. There would have to be reconnaissance. Peltz would need to learn not only what fortifications the Germans had in place at La Giorgetta, but also whether they had mined the hillside.

"Give me a plan that makes sense," Gofton-Salmond concluded. "And then, my boy, you want to fight, you'll get to fight."

TWELVE

Peltz moved through the darkness, listening to the silence. His face was blackened with charcoal, and it seemed as if the night sky had been painted with it, too. The moon, even the stars, were obliterated by heavy dark clouds. It was impossible to see more than a rifle's length in any direction.

Peltz's foot came down on a twig, and the noise was an explosion. He was certain a German patrol must have heard it. He came to a halt, weapon raised, waiting for them to charge toward him. But there was nothing. Just the thick, disorienting darkness, and the terrible realization that he was totally lost in enemy territory.

He tried to search for something familiar, but it was hopeless. The complete strangeness of it all, shadows layered upon shadows, was indecipherable.

The one certainty was his predicament. He needed to make his way back to camp before the sun rose and he was revealed to the Nazi snipers. Or a patrol stumbled on him. Or he stepped on a mine.

In another lifetime he had studied to be an engineer, and now his only hope, he decided, would be to apply a cool, mathematical logic to the situation. Peltz crouched down, his back against a tree for support, and spread his map out on the ground.

With one long thick finger, he began tracing a line on the map.

Shortly after eleven that night, he had left the low flat plain near the Fosso Vetro where three days earlier the German guns had fired at him. The countryside dipped a bit, and he had crossed a wide, mud-brown field packed with British mines. On the map, he re-created this journey with a smooth trace of his finger. But the reality had been more difficult.

Before he had headed out from camp, he had been briefed by an infuriatingly offhanded intelligence sergeant. It would be prudent to disregard the *X*'s indicating mines on the map, the man had warned. "Just guesswork, really. Best to assume the little buggers have been sprinkled about like handfuls of seeds," the sergeant said as if he were making a joke. But Peltz was not amused.

He had made his way through the broad field that night with an edgy caution, alert, his eyes focused on the ground. In the darkness every step was a drama, a moment of decision. And as he relived those tense moments, he realized where things must have gone wrong.

According to the map, once he got past the field his route should have been uphill and straight. He would enter a thick forest. When the trees began to thin, there would be a small farmhouse and a barn. A large wheatfield surrounded the homestead; undoubtedly it, too, was planted with mines, but these would be German. And as the terrain started to rise more steeply there was a cluster of stone and wood houses: La Giorgetta.

On the map it seemed quite direct; follow the compass arrow straight north. But put off-course by his circuitous path through the British minefield, he must have entered the forest from an odd angle. He was certain he had proceeded uphill. The compass proved he had been heading north. Yet where was the farm? Or the fields? Or the mines? Unless, he began to wonder, he had somehow passed the farm and he was already in the German minefield.

Peltz had a sudden sickening thought. Perhaps his map, splayed so carelessly on the ground, was concealing a German Teller mine. Trying to control his panic, he threw the thin sheet aside and, with just the very tips of his fingers, gently brushed the soft dirt. He waited for the sound of his nails scratching against metal, but there was only loose, cool earth. For a moment he took comfort in his simply being lost.

Yet when he looked back at his watch, another worry reclaimed his attention. It was three A.M. The Romagna dawn rose high and early. Without the cover of darkness, he would never make it back to camp. The Nazi snipers were infamous for their accuracy. A six-foot-three-inch target would not even give the bastards much sport. He looked again at the map. It told him nothing. He was in enemy territory, and he was lost.

He made up his mind: better to live and try again. He took the radio off his back and called the Brigade artillery.

Minutes later a phosphorus shell lit up the night with its unnatural dappled glow, and he got his bearings. He was nowhere near La Giorgetta. But he could see a way through the woods that would lead up toward the plain. And from there he could find the canal. When the artificial light faded, he started back.

It was just after dawn when an embarrassed Peltz went to the colonel and made his report. "Nerves, my boy. Nerves," Gofton-Salmond told him with an easy, uncritical laugh. "Get some sleep. Tonight's another night."

Night again. But it was not like the first night. The stars were no longer hiding. Peltz stood near the shallow irrigation ditch and looked up at the Italian sky. A pale sickle moon was above him. In the distance, he could make out a bucolic landscape of carefully delineated fields, a patchwork of contrasting dark shades spread across a softly moonlit hill.

Unprepared for this tranquil beauty, he found himself suddenly thinking of another countryside. And whether this same moon was shining down on Poland. On the fields of Zabiec. But he refused to let these thoughts take him any farther away. In a moment his mind was once again shuttered to any extraneous considerations. And he went forward, off into the first minefield.

Tonight he was all industry. On his long way back the previous evening, Peltz had realized that if a sizable force was going to cross the minefield, a path would need to be cleared. Defusing the mines would be the dicey work of the Pioneers; they were the specialists with what Peltz admiringly called "watchmaker's fingers." However, it was Peltz's job to identify the route, and he went about this tense work with discipline and attention.

Before he had left camp he had ripped several white bed sheets into long thin strips. Now he carefully tied these flags around the exposed metal handles of the British mines that would need to be defused. He was relieved when he was done and finally standing near the tree line.

Tonight, under the starry, moonlit sky, he had no difficulty finding the way through the trees. He made his way past the hardscrabble farm with its tin-roofed barn, and into the wheatfields. These fields were the only approach to La Giorgetta and, as the intelligence had anticipated, they had been mined by the Germans. A path wide enough for tanks would need to be marked. It was demanding work. And all the time the German troop compound at La Giorgetta was in sight; he could make out the dark, angular shapes of the buildings. He hoped the soldiers were asleep.

By the time he finished, a glance at his watch told him it was too late to continue on toward the compound. Dawn was approaching. He would return tomorrow night to locate the artillery and machine gun emplacements.

Exhausted, Peltz started back. As he was coming to the end of the field, he heard voices. He dropped at once to the ground, and crawled to a small dip in the terrain. He lay there, not moving, as the voices got louder.

There were two men and they were speaking German. One was loud, his words spoken in a disconcerting, almost feminine, high pitch. The other was more taciturn. He grumbled an occasional *"Ja."*

The two soldiers came straight toward Peltz. He heard their legs brushing against the high grass. He saw the curved outline of their helmets.

Peltz, his tommy gun in his hands, knew he could kill them. One sustained blast and there would be two dead Nazis. But if he did, another German patrol might hear the gunfire. And what about the bodies? Even if another patrol was not nearby, someone would come looking for these two in the morning. How long after that would it be before the path he had so carefully marked through the minefield was discovered? He had covered up the white strips with handfuls of dirt, but if the Germans became suspicious and searched the area, this small attempt at camouflage would fail. The enemy would discover the outlines of a corridor wide enough for tanks, and they would realize an attack was coming. They would be waiting, prepared in their bunkers on the high ground. His men would be slaughtered.

Peltz lay in the dirt. His left hand gripped the ribbed stock of the weapon. His index finger closed on the smooth metal trigger. In his mind he heard the gun exploding. It would take only a small assertion of will. But he did not fire. He held back, and the two Nazis walked by.

In camp, the colonel was furious. "If you have the opportunity to shoot the enemy," he lectured a startled Peltz, "do it! Don't think so much. That's what's wrong with you people. You think too much."

Peltz did not argue. He returned to his tent to rest for the night's mission. But he could not sleep. He kept replaying the moment when he could have killed two Nazis, and he kept examining his hesitation. Had he genuinely been concerned about jeopardizing the raid? Or was it something deeper, more intrinsic? Had the colonel, however crudely, intuitively focused on a more fundamental problem: Could a Jewish soldier find the courage to confront the Aryan supermen? Perhaps race was destiny. And perhaps his race was no match for the Nazis.

On the third night, Peltz made his way into the heart of the German camp. He moved noiselessly through the tall grass, excited by his own daring. He listened for a patrol, and when he did not hear anything, he darted out of the grass to a clump of trees. For a while, he waited among the trees, hiding in their shadows.

Across from him, rising up in the darkness, he saw a concrete bunker. He studied it and realized he would be able to inspect the entire compound from its flat roof. It would be an easy climb—a leap, really—to the top, but the bunker was probably manned. He wondered if he could get there without alerting the soldiers inside. Then he remembered how he had hesitated the night before, and that made up his mind.

Peltz ran from his cover and jumped onto the flat roof. He landed, he was certain, with the impact of a mortar shell, and he prepared to fire at the first soldier who came out to investigate. But no one emerged.

He lay there for hours, concentrating, determined to fix it all in his mind. The location of the machine gun nests. The unvarying routes the enemy guards patroled. The stone farm buildings filled with the rumble of nocturnal coughs, grunts, and wheezes, sounds that unmistakably identified them as troop barracks. He was the perfect spy. He memorized it all.

In time, a gramophone began to play below him in the bunker. He heard the churning sound as the machine was wound up, and then the hiss as the needle was placed on the grove in the record. Over and over, Marlene Dietrich sang.

> . . . So wollen wir uns wieder seh'n
> Bei der Lanterne wollen wir steh'n
> Wie einst Lili Marlen' . . .

The soldiers would join in too, a grave melancholic chorus.

> . . . Wenn sich die späten Nebel drehn
> Werd' ich bei der Lanterne steh'n
> Wie einst Lili Marlen' . . .

The mournful lyrics pierced the night. But Peltz did not let the music distract him. It was as if his brain had truly been divided into independently functioning lobes. One side was operational, focused on gathering the information for the report he would make to the colonel. The other, conducted a more personal sort of reconnaissance.

Listening to their voices, he began to assimilate another knowledge. For the first time, he had a sense of the enemy. If they could be so obsessive in their romanticism, so human in their melancholy, then what else were they but men? These were not supermen. They were soldiers who were lonely and unloved, and who would bleed when they were shot.

When Peltz returned to camp before dawn, the sad song was still playing in his mind.

THIRTEEN

Leah Pinchuk was hungry. She walked slowly behind her mother, the two of them wandering through the woods, searching for food. It had been nearly three days since they had fled their hiding place and run into the pine forest that bordered Reflovka.

Yesterday morning they had discovered some blueberry bushes. Leah had never liked blueberries. In past summers her mother would bake a blueberry pie and serve it with fresh cream, and she would refuse even a forkful. Yesterday the blueberries were a feast. When she could not eat any more, she had filled her pockets.

But she had finished the berries she had picked, and they had not found anything else. Leah licked her fingers for a taste of the sweet juice. The residue was all gone.

They kept walking. At one point Leah asked about her father. Her mother started to answer, but the words would not come out. Leah did not ask again.

There were no trails through the forest. Her eyes darted around hoping to find something they could eat. It was difficult to concentrate.

"What should we do, Mama?" They had been walking all day and

Leah was exhausted. She was a fourteen-year-old and she wanted her mother to have an answer. She wanted things to be normal.

Reitze looked for something familiar. The forest was immense. The canopy of branches let in very little light. Finally she said, "We're lost."

But Leah knew they were not lost. To be lost meant that you could not find where you were going. They had no destination. There was no place they could go.

If there was nowhere to go, then there was no longer any point in being alive. At the end of the forest, there had to be someplace for them.

"Leah," she heard her father asking, "what's your brother's address?"

"Sixteen Betzalel Street, Papa."

"My smart girl."

Now she could continue.

On the afternoon of the third day, it started to rain. The damp bracken flattened beneath their feet. Near a stand of birch trees they came to a path. They followed it. It was somewhere to go in the rain.

A farmer, leading two cows, was down the path. He was old and as thin as a scarecrow. Leah grabbed her mother's hand.

The farmer appraised the two women carefully. Their hair was wet from the rain and their clothes were disheveled. His eyes focused on the spot on Leah's dress where she had removed the yellow star. He did not say anything, but Leah knew what he was looking at. Leah wondered if she could find the strength to escape.

"You know what?" she heard the farmer say as if they were already in the midst of a conversation. "Come with me and I'll give you something to eat. I have a place where you can sleep."

The man spoke in Ukrainian. Leah knew the peasants hated the Jews. But the way he spoke, as though he were inviting neighbors over for dinner, was reassuring.

The farmer led the way, and they followed him. For the first time in days, Leah felt a lightness that was the beginning of hope.

* * *

Across from the farmer's house was a barn with a silo attached. He took them to the barn and told them he would return with food. When he left, they heard the sound of a wooden bolt dropping into place.

"Mama!" Leah cried.

"*Shah.* He wants to be sure no one will find us by accident."

Leah did not know if her mother believed this, or was simply trying to make her feel better. She sat on the dry, warm hay. The rain had stopped but it would soon be dark. She felt faint. Her throat was dry. What other choice did they have than to trust the farmer?

He returned with milk, bread, and pieces from a freshly cooked chicken. Leah ate ravenously. She had not eaten in so long that her stomach began to hurt. But she did not stop.

When the farmer left and bolted the door behind him, Leah was not concerned. She fell asleep filled with an admiration for the peasant who had helped them. He did not know her, but he had felt sympathy for her distress. Perhaps there could be a future.

A sudden pain woke Leah up. A Ukrainian policeman stood above her, jabbing the butt of his rifle into her back. "Mama," she cried as the weapon slammed into her again. "Help me!"

Her mother was held by two policemen. They tightened their grip as she struggled and cried. "What are you going to do with us?" she asked.

A policeman jerked Leah to her feet. Standing by the door of the barn was the farmer. She looked at him, and he smiled.

The Ukranians said they were taking them to the police station in Vladimritz. It was night, but Leah could see the road well enough to realize they were not going toward town. They were taking them to the woods. She knew what they did to Jewish women in the woods.

When she started to cry, a policeman hit her in the back of her neck with a rifle butt. She took her mother's hand and walked on. There was nothing else she could do.

But the thought of them hurting her, doing things to her, was unbearable. And she did not want to die. These were her last moments, and she found herself desperately wanting to live. An instant ago she had been willing to surrender, but now she found she could not.

"Mama," she whispered in Yiddish, "I'm going to run."

"Leah, no. Please. I could not bear to see your blood." Leah wanted to believe her mother would help her find a way out of this. But at the same time she knew there was no way through all this unless she found the will to create it.

"Mama, if I stay, you'll see my blood."

She kissed her mother on the cheek and then she ran.

The policemen shot at her. Leah heard the bullets flying by her in the night.

She ran, and when she reached a fence she somehow put both her hands on the top rail and pulled herself over. The rifle shots were very loud. She heard the policemen yelling, but she did not look back.

She stumbled into a wheat field, crouching low as she rushed on, trying to conceal herself among the tall stalks. Her shoe came off, but she did not stop. The shooting continued. She kept down, racing through the field, heading toward the forest.

FOURTEEN

When Carmi and Peltz had served together in the Palestinian Police, they had seen a man blown up. A British officer had led the way into an Arab building in the village of Dabburiya. As soon as he opened the door, an explosion blew the man to pieces. Both Carmi and Peltz had been close enough to need to wipe his blood from their faces.

This image was in Peltz's mind as he planned his attack on La Giorgetta. During his nights of reconnaissance, he had discovered that the Germans had seeded the muddy approach to the hilltop compound with three different kinds of mines.

The ones with their three-prong detonators sitting above the dirt were the *Schutzenminen*, or "Bouncing Betties" as they were more often called. Beneath the ground was a metal canister not much bigger than a soup can containing 360 hard steel balls. When a combat boot stepped on one of the prongs, there were in rapid succession two explosions. The first flung the mine up in the air waist high. An instant later, the second spewed out the steel. The mine was not designed to kill. It was a torturer's weapon. The flying metal could tear off a leg above the knee.

The Schu mine was less ingenious but more deadly. It was a wooden box slightly larger than a bar of soap and packed with a quarter pound of TNT. The pressure of a soldier's foot drove a nail into a detonator and the explosion could butcher a man.

The bigger mines, the size and curious shape of two dinner plates placed rim to rim, were Tellers. Their long five second fuses and pounds of explosives were designed to incapacitate slow-moving armor-plated tanks. However, Peltz knew the Tellers had also proved quite effective against advancing, tightly clustered infantry. A single explosion might leave a squad of bloodied, screaming men on the ground.

The mines would need to be neutralized before the attack. It was crucial. But how? Peltz worried.

The simplest, least dangerous method would be to hit the wheat field with a sustained and concentrated artillery barrage. But that would signal to the Germans that an attack was coming. Without the element of surprise, his charge would be doomed.

Sappers from the Pioneer battalion would have to go in first, Peltz reluctantly conceded. Perhaps they could simply lift the mines from the ground and create a corridor wide enough for the men and tanks to advance. Yet as soon as he thought this through, he knew the Germans would notice the path. They would be waiting for an attack.

In the end, he realized there was only one way to neutralize the field—and it was the way he had hoped to avoid. The mines would need to be disarmed by hand.

The fuse in the top of the metal casing would need to be removed. It was an act as simple as unscrewing a light bulb from its socket. But if there was a clumsy twist, an inadvertent amount of pressure, or even an awkwardly aligned groove, hundreds of brittle pieces of hard-driving steel would rip through the sapper's face.

And the disarming would need to be done at night. In enemy territory. With a company of Wehrmacht troops sleeping not more than five hundred meters away.

Resigned that there were no other options, he went to tell Haim Brot. Brot was the Pioneer platoon corporal who led the sappers. He was also Peltz's friend. They had served together in the south Dead

Sea region. But this did not make his decision any easier; Peltz knew it was unforgiveable to ask this sort of thing from anyone.

Brot listened to the plan and had one concern. This time of year the detectors were of no use. There was too much mud. His men would have to locate the mines with their bayonets.

"Then that's what you'll have to do," Peltz ordered flatly.

Peltz waited in his tent that afternoon. He lay restlessly on his cot until it would be time to lead Brot and his squad across the dark Italian countryside. Unable to sleep, he found himself thinking about a night he had shared with Brot years ago at the police post near the Dead Sea.

Peltz had recently returned to Sdom from two weeks of medical treatment for his leg. Before Peltz left Jerusalem, a doctor had given the wounded hero a recording of a vocal quartet singing Brahms's arrangement of Hungarian folk songs as a present. Peltz was a sentimental man, and he was moved by the gift. He played the record often.

One night Brot came by to visit. It was a pleasant, almost cool evening, and the gramophone was by an open window of the tin-roofed police shack. The men listened for hours.

Peltz's memory of the evening was still vivid. He heard the music: Brahms's songs, their melodies carried by breezes blowing over an ancient biblical sea, traveling off into the vast gray desert, heading out in the empty night to the Arab camps.

As he lay on his cot, one recollected melody eventually tapped into another. Brahms trailed into a sad, wistful soldiers' chorus of "Lili Marlen'." It was as if they were movements in a single piece. And a connection was made in Peltz's mind. He understood its significance. The hero of Sdom had fought the Arabs and won. He would do the same against the Nazis.

It was time. He got up from his cot and went to get Brot.

FIFTEEN

Two hours later, Peltz was lying in the mud near a busy, methodical Brot. Along with a half-dozen other men, they were in the field below La Giorgetta. Peltz had a tommy gun in his hands, but he knew there was really nothing he could do except watch. It was all up to Brot and his team.

For Peltz, the instant when the tip of Brot's bayonet blade prodded into the ground was always the worst. If a concealed trip wire was still attached to the mine; if the mine had been accidentally buried on its side, its pressure plate inverted beneath the dirt; if for any of too many reasons he had no trouble imagining—fatigue, poor night vision, carelessness—Peltz knew what would happen. The mine would explode. And both of them would be maimed. Or killed.

But how could Brot, the bayonet held high in his hands, know what was waiting beneath the dark, thick mud? The suspense was terrifying. Peltz imagined that the Germans in their distant bunkers could hear the pounding of his racing heart.

For three hours Peltz stayed near Brot in the moonlit minefield. He was in awe of Brot's composure.

When Brot finally announced that a path wide enough for the tanks had been defused, Peltz led the way back to the forest. As soon as they were surrounded by the high trees, Peltz ordered the squad to halt. Parched, he gulped down a canteen of water, and then finished Brot's too. Still, he could not wash the taste of fear from his throat.

"Tomorrow morning it is then," Col. Gofton-Salmond announced when Peltz, less than a hour after leading the Pioneers back, had finished his report. "At ten."

"*Morning,* sir?" Peltz questioned, unable to disguise his surprise.

He had been taught it would be a mistake for a raiding party to attack a fortified camp in daylight. Carmi's company had challenged an enemy position dug in on the other side of a fast-moving creek one afternoon, and the charge had turned into a disaster. They were pounded by 81mm mortars and shredded by a hail of light machine gun bullets. Six men were killed, and more were wounded. It was a small miracle that they were able to make an orderly withdrawal without further loses. Within the week, the British company commander who had given the order to advance in daylight had been removed. Peltz could see his own mission ending in a similar defeat.

But Gofton-Salmond was never peremptory; and he was fond of the young officer. "Last thing in the world the Jerries will expect is an attack in the daytime. Move out smartly, and you'll be past their artillery before they realize what's happening."

As Peltz thought it through, he began to appreciate the colonel's strategy. Day or night, the charge into La Giorgetta would be an adventure. But if his force could make it to the forest before the German eighty-eights had an opportunity to rough them up in the open land near Fosso Vetro, one of the enemy's great advantages would be nullified. And his force had an element of support Carmi's squad had lacked—tanks. He would have the monster four-track Churchills, their cannons firing, leading the way through the wide gap now cleared of mines, and into the heart of the enemy compound.

"Yes, sir," Peltz agreed. "We go at ten."

"Good," said the colonel. "Now get some sleep."

Drained after the tense night, it took all of his energy just to salute. The colonel saluted in return. But as he walked Peltz out of the command post, he put an arm around the young captain's shoulder. "You go out and kill them all tomorrow. Then come back here, and we'll have a nice cup of tea."

Haim Brot was waiting for Peltz outside his tent, and at once Peltz knew something was very wrong.

"I've been thinking it over," Brot said before Peltz could sit down. "I think I might've missed a mine. Maybe two. I got confused. The night . . ." His voice trailed off.

Brot was pacing back and forth as he talked. All his previous calm might never have existed.

"What are we going to do, Johanan?" he demanded.

"I don't know." Peltz felt exhausted to the point of despair.

"I could go in ahead of the tanks. Clear the way for them myself," Brot brashly suggested.

Peltz imagined his friend leading the charge, defusing mines under enemy fire. Then he said, "I'll talk to the colonel. Postpone things."

But as soon as he spoke he realized how it would look to Command. "You people," he heard Gofton-Salmond sneering. It would be another lost opportunity, another disgrace. Jews would once again be backing down from the Aryan supermen.

Yet what choice did he have? Could he order tanks, his men, to charge through a minefield?

"What are we going to do?" Brot repeated.

Peltz had no answer. So much was at stake.

Finally, Peltz decided he needed to sleep. After he was rested, he would be able to shut out all the extraneous feelings. He would have the objectivity to reach a soldier's well-reasoned decision. "Come back tonight," he told his friend. "We decide to call it off, there's still time. For now, let me get to bed."

But when Brot left, Peltz found that it was impossible to escape into sleep.

* * *

That afternoon when he went to the mess tent, Peltz's mind was in turmoil. He had still not made a decision. He sat at a long wooden table, and soon Arie Pinchuk joined him.

"Have you heard what the wireless operator picked up?" Pinchuk asked.

Peltz was locked into his own thoughts. He did not want company, and he ignored the question.

But Pinchuk was too agitated to be put off. "The Germans know we're here. In Italy. Know what they called us? The Jewish *Plattfuss* brigade."

"The flatfoot brigade," Peltz quickly translated. Once spoken, the words cut through his mood. "What else did the bastards have to say, Arie?"

"Garbage. Who can remember?"

"Try," Peltz ordered.

Pinchuk considered, and then repeated a version of the broadcast that, he decided, was suprisingly accurate: "It is sad to watch how much the English people have deteriorated that they have sent a Jewish flatfoot brigade to fight in Italy against an Aryan army. But the German people can be assured that the Jews will never fight. They will run at the sight of our Aryan Wehrmacht."

Peltz rose abruptly from the table.

"Johanan, what's wrong?" Pinchuk asked.

But Peltz did not bother to answer. He hurried out of the tent to look for Haim Brot.

SIXTEEN

It was just before ten on a sunny late March morning when ninety-six men, B Company of the Third Battalion reinforced by a platoon from D Company, all under the command of Capt. Johanan Peltz, assembled at Fosso Vetro. Peltz was no longer on edge. He was ready.

He raised his right arm high into the air. As he did, he looked back over his shoulder. He saw a line of combat soldiers, the Star of David on their uniforms. A Jewish army was assembled for its first battle in two thousand years. In Palestine, Peltz had skeptically dismissed all the Zionists' lofty talk. But at that moment in a field in Italy, the idea of a Jewish nation suddenly became real to him.

Peltz brought his arm down. "Forward!" he ordered.

The Germans were caught by surprise. It was not until the force was nearly at the tree line that the German guns began their barrage.

By then it was too late. The troops, advancing rapidly, had crossed the open ground without casualties.

Three Churchill tanks, driven by men from the Royal Irish Horse Regiment, were waiting as planned at the rendezvous point in the forest. "Right behind you, mates," a tanker called from the turret. Peltz saluted him, and led the way.

Despite all his reconnaissance, this was the first time Peltz had been in the forest in daylight. He felt as if a blindfold had been removed. He had no difficulty finding the path uphill to the ramshackle farm.

The pounding of the German mortars was growing louder, but he could also hear the rumbling of the three tanks behind him. The sound of the big machines smashing their heavy way through the forest, unstoppable, their 75mm cannons ready to fire, was reassuring. Peltz's mind was racing, alert and intense to every new moment. But as he moved closer to the minefields he could not help thinking that the attack was going precisely as planned.

The wide approach to La Giorgetta was a jumble of menacing noise. The Spandaus sprayed the air with a high, speeding whine. The deep thuds from the mortars were more intermittent, an oddly rhythmic counterpoint. And through it all there was the crack of rifle fire: sharp, ominous, precise.

Peltz ordered the tanks, along with most of the force, to stay back until it was certain that all the mines had been cleared. They assembled in a line in the tall grass that bordered the field, firing without accuracy or commitment toward the unseen German positions. The battle had begun, but for now they would have to wait.

Peltz charged ahead into the minefield. He led a platoon of men, including Haim Brot.

Near the end of the field, about three hundred meters from the compound of stone buildings the Germans occupied, was a wide ditch. The improvised plan was to provide cover fire for Brot while he neutralized whatever mines remained, and at the same time advance toward this bit of sheltered terrain.

As soon as Peltz moved out into the open and into the full force of the enemy fire, he was overwhelmed. He was running, shooting, for a time even crawling toward the ditch. It was all a wild blur.

But there were moments when, like scenes flashing through the window of a train emerging from a tunnel, brief distinct images came into focus. Brot, just five yards away, on his knees as if in prayer, fingers working the fuse of a mine. A radiant glow from a bunker, and in the midst of all the confusion, Peltz, on his belly, taking aim. Brot, short legs taking careful measured strides, the prodding bayonet in his hand, as he approached another mine. A triumphant yell: "The last one. I got it, Johanan." And Peltz rising tall from his defensive crouch, turning first to the tanks to order them to advance; and then to Brot.

Through the tumult of noise, he heard the lumbering machines move forward. He could see Brot, his bayonet high in the air as he, too, signaled to the tanks.

And in that instant: across Brot's chest, from his shoulder to his waist, a thin red diagonal as straight as if it were drawn with a ruler appeared.

"Stretcher bearers," Peltz yelled from somewhere so deep inside him that it seemed to climb over all the other sounds. The chaos was immediately slowed to the point of entropy. The blur vanished. And order, as tight and as terrible as a prison, was restored.

Peltz raced to Brot and gently raised him up from the ground. Peltz held him in his arms, wanting to comfort him. But it was pointless. His friend was dead. Peltz felt as if everything had been lost.

A stretcher bearer began tugging at his friend's body; and with this intrusion, Peltz regained perspective. He tightened his control.

Peltz lifted Haim by the chest. His arms were wrapped across a battle shirt wet with blood. The medic took the feet. They lowered the dead man onto a canvas stretcher.

There was nothing else that could be done.

Charging as if through the flames of a raging fire, Peltz ran toward the ditch, and La Giorgetta.

In one sustained dash, he made it. Most of the two platoons, Peltz observed after a quick inspection, were intact. They had lost some men, others were wounded. But those that were with him, those that were crowded up against the dirt walls of this hollow trench, were ready. On his command they would go forward.

Peltz looked over the embankment toward La Giorgetta. Substantial puffs of bright smoke hung in the air outside the pale stone buildings. It was as if the soldiers inside were breathing fire at his men.

He looked back over his shoulder to locate the tanks, hoping that by now they were in position to support the charge. And he saw Corporal Reccanaty of the stretcher bearers, a big man like himself so there was no mistaking who it was, dumping Haim Brot's body onto the muddy ground.

"You bloody bastard. I'm going to kill you!" Peltz shouted above the roar of battle. His body was shaking. His friend was a hero, and this insult was unforgivable.

But he held back. There was no point in going back. Haim was dead. The stretcher bearers were ordered to carry off only the wounded. Reccanaty was not the enemy. The path to honor lay in only one direction.

Peltz was still seething, but now his tremendous anger was focused. It was so pure it was almost calming. He stood and faced his men. His back was to the German guns as though he were invincible.

"Fix bayonets!" he ordered.

There was the distinct sound of the steel bayonets clicking onto the steel barrels of the rifles.

"Up . . ."

Two platoons of soldiers rose.

". . . and *charge*!"

Shouting madly, a roar in a Babel of languages and accents, Hebrew, Polish, Yiddish, English, the troops raced forward, bayonets in their hands.

It was nearly three hundred meters to the German positions. The air was dense with bullets, the snorting of the mortars, the lashing of the Spandaus. Peltz felt as if the Fatherland were firing all its remaining ammunition into this one field, determined to stop this one charge.

But they kept running forward, shouting incoherently. Peltz could hear the bullets. He was taking long, measured strides. He gripped his rifle as if it were the weight that would hold him to this world. And all the time he kept getting closer, closer to the stone buildings of La Giorgetta.

Still running, he saw a machine gun barrel extended like the snout of an evil animal from a trench.

"Uzi," he shouted to Sgt. Uzieli, "To your left."

Without breaking stride, Uzieli tossed a grenade toward the nest. As it exploded, the men following him sprayed the smoke with rifle fire.

And the charge continued, loud, forceful, and pounding.

All at once they were in the midst of the enemy compound. The main force directed its fire on the outer circle of stone buildings. But Peltz moved on and about a dozen soldiers followed. He led them to the big farmhouse he had identified as the headquarters of the German command.

"The door," he yelled to the men.

As they smashed down the door, Peltz, in one long leap, bayonet extended in front of him like a lance, crashed through a wide double window. The glass shattered and Peltz, driven by an unstoppable momentum, propelled himself across the room. A German officer, a short thick man with a look of startled terror in his eyes, tried to raise his Luger. Peltz ran him through with his bayonet.

The man was driven back to the wall, impaled. Peltz twisted the blade in the German's gut as though it were a key in a stubborn lock, but he could not pull the tip free from the wall. He kept trying, but it was futile. The German officer, his last moment of atavistic fear frozen on his pudgy face, remained pinned to the wall like a specimen in a display case.

Peltz left him there.

He drew his revolver before he entered the front room. It was unnecessary. There were eight Germans. Two had been bayoneted. The others had been shot. His men had killed them all.

When Peltz emerged from the headquarters building, he heard only sporadic gunfire. The enemy force had retreated across the Senio. Those that had remained in La Giorgetta were dead, or too badly wounded to escape.

Peltz saw Sgt. Moshe Magdid standing with some men by a long, narrow building. On his night of reconnaissance, Peltz had determined it was the barracks.

He ran over quickly. "You checked it out?"

"Yes, sir," said Magdid. "Empty."

"The cellar?"

There was an awkward silence. "Give me your tommy gun," Peltz ordered.

The sergeant handed him the weapon, and Peltz went in the door. He looked and saw the room was empty. Then he searched for the stairwell to the cellar.

He threw a grenade into the dark subterranean room, and before the smoke cleared he went in shooting. The ceiling was so low that it was impossible for him to stand, and he had to fire in a crouch. He was bent over and he was spraying the tight, confined space with bullets.

He finally stopped. It was dark, but he could see that a thin red stream was running over his boots. He thought it might be blood. Suddenly he was very dizzy. An unsettling rush invaded his head, and he feared he had been wounded, and it was his own blood flooding the room.

It took all his concentration to pull himself up from his crouch, and to stagger up the steps leading back to the barracks.

The sergeant was waiting for him. He took one look at Peltz, breathed in the ripe, pleasant smell, and announced, "Sir, you are drunk."

As canteens of revitalizing water were poured over Peltz's head, soldiers confirmed that only several large barrels of wine had been hiding in the cellar. And Peltz had shot them full of holes.

"I imagine," said a grinning Peltz, "I'm the only company commander who ever got drunk during a bayonet charge."

An hour later, Peltz gave the order to pull out. A small force would remain to hold La Giorgetta, and he would lead the rest of the men back to camp.

As the men assembled, Hyman Zekowitz approached Peltz.

"With your permission, sir," he said.

Peltz watched the short man unbutton his tunic. Pressed against his chest was a piece of cloth. As Zekowitz unfolded it, Peltz saw it was a flag with two blue stripes and a Star of David in the center.

"By all means," said Peltz.

Zekowitz attached the flag to the barrel of a rifle, and then climbed to the roof of the German command building. The butt of the rifle was firmly tied to the chimney. High in the Italian sky, the Star of David blew in the small, soft breeze.

Standing on farmland littered with the bodies of dead Wehrmacht soldiers, the men cheered.

They had won. Jews had beaten Germans.

It was getting dark by the time Peltz had crossed back over the minefield and searched for the field hospital. It had been set up in the forest, and as soon as he entered he was given the casualty report. Nineteen men had been killed. Twelve had been wounded. He asked to see the bodies.

They directed him to the makeshift morgue, but he could not find Haim Brot.

"He must be still out there," he was told. "We'll find him in the morning."

Peltz thought about his friend out there alone in the night. "I'll get him," he said.

It was a long search. The field had been savagely disfigured, ripped and torn by so many shells, rutted by the tanks. But after about an hour he found the body.

He closed Brot's eyes. Then he picked his friend up. Peltz placed him over his shoulder with the care and concern one would give a child gone down for a nap, and began the long walk back in the twilight across the minefield. In Peltz's mind, Brahms played the entire way.

SEVENTEEN

High in the Apennine foothills, Carmi stood daringly close to the edge of a narrow I-shaped rock ledge. A careless step, and it would be at least a hundred-foot fall. Yet he would not step back.

Behind him, smiling with genial unconcern, was a short man, skin the color of weak tea, with a shaved head and a broad sharp-edged sword called a *kukari* at his waist. He had led Carmi out onto this thin promontory because, he said, it offered an outlook on the entire sector the Brigade would be taking over tomorrow.

Carmi had to concede that the long meandering front line was effectively displayed. In the clear spring daylight the view was spectacular: the Senio, sinuous and shiny like a fat dark snake sunning itself in a garden, and beyond the river were hills and farmland, a placid checkerboard of soft brown and green shades.

Nevertheless, Carmi could not help wondering whether the tacit challenge was deliberate. He remained rooted at the tapered point of the ledge, determined to demonstrate that a Jew from Palestine could be as coolly oblivious to dizzying heights as a Gurkha raised in the mountains of Nepal.

Perhaps, Carmi suspected, the Gurkha officer was annoyed by his having to give Carmi a tour. After all, Carmi was only a sergeant. Normally he should not have been sent with the staff officers assigned to inspect the new front.

Yet although he had only middling rank in the British army, Carmi had important long-term responsibilities. Whatever Carmi learned in the course of this war was knowledge the Yishuv would draw on in the future. The Jewish officers understood that, and made sure Carmi was assigned to accompany the inspection team.

When Sgt. Carmi, feigning a breezy curiosity, asked the Gurkha officers about the plan for the final offensive, he was storing away information about the logistics necessary for a coordinated attack—artillery, armor, infantry, and perhaps even air—that would one day have other practical applications. And when in his fleeting exchanges with the Gurkhas, he took the measure of the ramrod-stiff men, appraising soldiers whose reputation for toughness and tenacity in battle was celebrated, he was silently plotting how these qualities might be instilled in another army. The staff officers were engaged in one reconnaissance mission, while the sergeant, with their support, was collecting information for another.

At last, to Carmi's great relief, the Gurkha officer backed off from the ledge and led them toward the flat plain. But he stopped long enough to flash Carmi a sly smile. "Well done, Sergeant."

Carmi, however, was not proud. All the long way down the rock-strewn hill, he fumed. He was angry at himself. It was a child's game he had foolishly played on the ledge. Why should he have taken the risk?

He was serving in a very demanding army. He had to be stronger than the enemy. And braver than the allies. He had to fight one war in a foreign land and then return home to fight another inevitable war. With an uncharacteristic bitterness, he decided this was the always tenuous Jewish condition. *The chosen people?* he felt like screaming. Chosen for what?

The short man nimbly led the way over the rocks and Carmi, following, allowed his mood to unravel further. Perhaps, Carmi thought playfully, he would be better off in some land like Nepal. He could farm in peace on a mountaintop. Or why not simply stay here?

The Italian countryside was tinted with the first shades of spring; he could only imagine it in the full rich bloom of summer.

But these thoughts, he knew, were only momentary, self-indulgent musings. His destiny was in Palestine. His life's work was to build a homeland for the Jewish people.

Before he had left for Italy he had exchanged farewell gifts with his wife. To their surprise, they had given each other the same book. It was *With and Without You*, a slim collection of poems by K. Semyonov. Carmi had learned one of them by heart. "Wait for me and I shall return," it began. "But wait for me a lot . . ." Carmi would return. And Tonka and their daughter and the land, the *eretz*, would be there waiting for him. He would return to them, and to their common future.

By the time he joined the other officers from the Brigade at the bottom of the hill, all his previous carpings had been shut out from his mind. And something had been restored: He had no doubts about what he had been chosen for.

Later that warm morning Carmi toured the trenches the Brigade would be taking over. They would need to be deepened. Jews, he pointedly informed the Gurkha officer who had him nearly dangling over the ledge, were a lot taller than his men.

The next day the Brigade moved into this narrow, hilly valley divided by the Senio River. They were ordered to wait at this position until the offensive began. In the meantime they were to hold the front line at all costs.

EIGHTEEN

CONFIDENTIAL

JEWISH INF BDE GP INT SUMMARY No.2 Dated 22 Mar 45

1. ENEMY SITUATION

The most striking difference to be noted regarding the new Bde front compared with our present sector (apart from totally changed topographical features) lies in the quality of the enemy tps which will be opposing us. Instead of having the 42 Jg Div as opponents, with its strong Austrian elements of doubtful morale and questionable Nazi sympathies, we will have to deal with the 4 Para Div representing the best German combat units. Not only has this div the newest and most abundant material, but, as important, its personnel are all specially selected men whose political loyalties as "good" Germans are unquestionable.

When Peltz read this memo, he had one thought, and he confidently went off to share it right away with Col. Gofton-Salmond. After all, the colonel had always supported his initiative.

The colonel listened with interest to Peltz's new plan. "You want to bring in some Jerries, go right ahead," he agreed. "Intelligence will jump at the chance to hear firsthand what the Wehrmacht is up to. Tongue snatching, we call it."

Peltz did not tell the colonel what he called it. The word in Hebrew was *nekama*: revenge.

For three nights Peltz led patrols into enemy territory, but they were uneventful. Moving in the darkness through difficult, rocky terrain, his men never encountered any Germans. It was very disappointing.

Peltz began to personalize the failure: The Germans were deliberately running from Capt. Johanan Peltz. To him, it was almost as if it were a competition. Peltz, the dogged hunter, on the trail of the scurrying Nazis.

Each night he went out, and each night he came back empty-handed. With each new failure, he became increasingly preoccupied with his quest.

"Peltz's manhunt," as it was known, became the talk of the mess. So when Maj. Storr received word from air reconnaissance that a German platoon directly across the river was withdrawing, he sent a runner to Peltz's tent. Retreats, the major believed, were the best time to grab prisoners. "Bound to catch Jerry with his pants down," he would say.

After another long futile night, Peltz was asleep. But on hearing the news, he quickly grew alert. Minutes later he reported to headquarters.

When he arrived, Peltz discovered that a patrol had already been assembled. Worse, it was under the command of Lt. Tony van Gelder, an English Jew who had transferred into the Brigade. He was short and had a tough, cockney feistiness. The troops called him "Mickey Rooney."

"Thought you might want to come along nevertheless," Storr told Peltz. "Your men would never be up and ready in time. Just remember, Captain—it's van Gelder's show."

A half-hour later a patrol of eight men and two officers moved in bright daylight into enemy territory. If the intelligence report was inaccurate, the snipers would still be at their posts. Peltz had seen their handiwork; they liked to aim for the unprotected Adam's apple. In the morning light, it would be as challenging as target practice. He moved forward in a crouch.

But there was no enemy fire. With the air reconnaissance map as a guide, they moved on. When the bunker was in sight, van Gelder ordered the men to take up positions near a stand of trees.

Van Gelder took a long look at the squat concrete structure and then handed his binoculars to Peltz.

"Nothing," said Peltz. There was not even a guard by the door.

"Intel was probably right," van Gelder agreed. "They moved out."

Peltz was disappointed. "We're too late."

"Only one way to know for sure."

Van Gelder moved the men out toward the bunker. They proceeded cautiously; the sound was muffled, yet distinct. If the enemy was inside, Peltz was certain that they would hear them coming.

They reached the entrance. Van Gelder motioned for his men to take positions on either side. Then he eased the door open and led the way in.

Fourteen German soldiers were sleeping in their bunks. The men from the Jewish Brigade did not move. They did not speak. They stared at the sleeping enemy.

Suddenly Cpl. Joram Lewy shouted in German: *"Heraus, ihr Schwein. Die Juden sind hier."* Get out, you pigs. The Jews have come.

The fourteen German soldiers woke up to see troops with the Star of David on their uniforms aiming their weapons at them.

Van Gelder ordered the Germans into a line and the men led them outside the bunker. The captives immediately understood their predicament. "I am not a Nazi. I'm a Social Democrat," one man pleaded.

Peltz was moved by what he was witnessing. Germans, the tormentors of his race, had their hands high in the air, helpless.

"I say we kill the bastards," said Lewy. He had been born in the Ukraine and raised in Germany before he immigrated to Palestine. He did not know what had happened to his mother, to his father, but he had spent many difficult hours imagining.

Some of the men agreed with the corporal. They waved their weapons at the prisoners, and taunted them in German. "Master race," they mocked. The prisoner who claimed to be a Social Democrat started to cry when a rifle was pressed against his temple. It was all rushing out of control, turning wild.

To his surprise, Peltz did not stop the men. In all his nights on patrol he had never thought about killing helpless prisoners. He had wanted to capture them. He wanted to prove that he was better, stronger, braver than the enemy. That would be his revenge on these "good Germans." But when he saw these Wehrmacht troops, their gray uniforms with their swastikas, he was consumed with an almost uncontrollable rage. A pure, visceral hatred surged through him. The image of the German officer he had left pinned to the farmhouse wall with his bayonet rose up in his mind, and he embraced it. If the men wanted to kill these prisoners, he was not going to intervene.

"Stop it," van Gelder shouted. "Enough."

Lewy still had his weapon pointed at a prisoner.

"Corporal . . ." van Gelder warned.

Lewy stared at his superior and finally lowered his rifle.

"They're needed for questioning," van Gelder said. "You want to kill the bastards, kill 'em in battle."

The men led the prisoners back to camp. Peltz, still shaken by the passions he had found himself caught up in, was glad that a confrontation had been avoided. But he also suspected something powerful had only been postponed, not banished. And when they were crossing back over the Senio and heavy enemy artillery killed two of the prisoners, he had no regrets.

HQ JEWISH INF BDE GP
Subject: *Prisoners*

1. I want you to impress upon all ranks the supreme importance of capturing live German prisoners, and of their being sent back quickly for interrogation through the proper channels.

2. I fully realize that there are a large number of men in the Jewish Bde Gp who have every personal justification for desiring to revenge themselves upon the Germans, and I am afraid that this may, in some cases, lead them to decide to kill every German they can, rather than take prisoners. This is a very short-sighted policy. Our object is to do everything in our power to hasten the defeat of the enemy and it has been proved time and time again that far more is gained by taking prisoners from whom information can be extracted under interrogation than by killing the enemy out of hand.

3. I want to stress one further point. However great the crimes which the Germans have committed against international and moral law, I am determined that the Jewish Brigade Gp shall act correctly in accordance with recognized conventions.

E. F. Benjamin
Brig.
Comd.

Later that week, Carmi was on night guard with his platoon near the shore of the Senio when a large German patrol attempted to infiltrate the Allied line. The water was low at this bend in the river, and he could hear the sounds of men wading across. They were trying to be quiet, but their deliberateness served only to amplify every soft, splashing noise.

"They're coming," someone whispered.

"Hold your fire," Carmi ordered.

The men waited and finally saw a line of indistinct shapes moving toward them in the night. Carmi raised his tommy gun and started firing. At this signal, his men opened up too.

The Germans continued to advance. Carmi could see the bright muzzles of their weapons lightening up the darkness when they fired. There were too many German guns; his men were greatly outnumbered.

A grenade exploded near Carmi. He heard a scream. His men hurled grenades back at the enemy. The two opposing forces were so close that in moments they would be fighting hand-to-hand.

Carmi got on the radio to company headquarters. "We need artillery support," he yelled.

He was told that it was too dangerous. His platoon would be in as much jeopardy from the Vickers as the Germans.

Carmi realized this. But without artillery his small group of men were certain to be overrun. He could see the Germans moving closer.

The man on the other end of the radio was Haim Laskov, and Carmi knew him well enough to trust him with his life. Improvising quickly, Carmi told him his plan: The men were going to raise their helmets up into the night sky. If Laskov heard the Vickers smashing into steel, he should adjust the range.

"Fire!" he told Laskov.

The sound of bullets hitting against steel rang out in a harsh tattoo.

"Too close. Too close," Carmi screamed.

"Israel . . ." Laskov tried.

"Again!" Carmi insisted. "Fire."

The Vickers spewed hundreds of bullets directly into the German advance.

"Again!" Carmi yelled.

Soon Carmi reported that the enemy force was pulling back. He kept his platoon in position for the entire night in case the Germans returned. They did not, and Carmi realized he had learned a valuable lesson about the effectiveness of short range over-the-head fire. He stored it away.

In the morning's first light, Carmi sent Moshe Zilberberg out into the field to search for wounded men. He carried a flag with the Red Cross on it. When he was not serving as medic, Zilberberg was also the company's barber and cantor. He possessed a clear, firm tenor; to Carmi's ear, it was a voice that was spiritual yet unself-conscious.

Carmi had his back to the field, but he turned when he heard three quick shots. A German sniper had targeted the medic. When Zilberberg's body, a bullet wound just below his heart, was carried back to the trenches, Carmi tried to calm himself by finding the lesson in this experience, too. Never, he told himself, trust the enemy.

It would be another small wisdom he would take with him into a lifetime of wars.

Before Pinchuk had the opportunity to fire at the enemy, he was removed from the line. The battalion motor transport officer had been transferred and Col. Gofton-Salmond promoted Pinchuk to captain to take his place. He tried to turn down the job and the promotion, but the colonel would not allow it.

Pinchuk was filled with despair. The war, Pinchuk now realized, would be fought by other people. Men like Peltz and Carmi. He commanded trucks, a captain of invoices. What use was he to the Yishuv? To the Brigade?

And if he could not fight, if he could not be in the front lines charging the Nazi defenses, no matter how the war ended, he had lost. How could he ever expect to erase his betrayal of his entire family? It would be impossible.

What did it matter, he finally told himself. It was already too late. Whatever few opportunities his family might have had were gone. They had lost, too.

NINETEEN

Once again, Leah Pinchuk walked through the forest. But now her mother was not with her, and it was night. She had lost a shoe when she ran from the police, so she undid the laces of the remaining one and kicked it off. It was easier to walk on the soft ground in her stocking feet.

She went on. Finally she found a flat patch of ground beneath a pine tree, and lay down across the scattered needles.

Why I am being punished? she asked herself. What did I do? I'll keep the Sabbath. I'll fulfill all the mitzvahs. Just please God, please God, save me.

But Leah suspected it would not be that easy. The memory of what she had seen in Reflovka made it difficult to place much hope in God's mercy.

Leah was on her own, and she had no idea what she would do.

She awoke at daybreak and saw a well-trodden trail that twisted through the trees. She decided to follow it.

It was the smallest of acts, as much a reflex as a choice, but it filled her with a sense of purpose. She was still frightened, but she had found the power to do something and it was reassuring.

All the while, she searched for a plan that would allow her to survive. The process was not always rational; it was easy to become tangled in wishful escapes. In her mind, they all ended with her arriving triumphantly at her brother's doorstep in Palestine. But as she worked her way through all the improbabilities, she came upon a strategy that seemed to make sense. She would look for Boris.

In the last sad days of the ghetto, Meir Pinchuk refused to have his possessions clawed at and fought over by scavengers. Instead, he chose to bestow them like a gift. It was not much, the cow, a favorite lamp, a set of heirloom china plates, but it was his fortune. In a gesture that was both generous and defiant, he gave it all to Boris Savchok.

Boris was a Ukrainian peasant who had helped out with chores in the Pinchuk house. He lived near the Satir River, about twenty kilometers from Reflovka, in the small farm village of Sofachov. When he came to town, he would stay with the Pinchuks. He would eat at their table. He would sleep in their bed. He was, Meir wanted to believe, a friend of the family.

Boris was also a Communist. When the Poles controlled the territory, they had imprisoned him. Meir had no sympathies for Boris's way of looking at the world, but he felt that in their common victimization they shared a bond.

Leah, too, had other reasons for looking hopefully to Boris. He was a tall, broad-shouldered man, heavily muscled, with brown sympathetic eyes and shiny waves of dark hair. If she were to use her imagination to construct a hero, he would resemble Boris.

It was settled in her mind. Somehow she would find a route through the forest. Then she would follow the river downstream to the town where Boris lived.

At the end of the dirt trail, there was a stream. A wooden bridge wide enough for a horse cart spanned the water. On the other side of the bridge were two large wooden structures, the homes apparently of prosperous families. Both had the same boxy shape, the same high pitched roofs, and red barns. One stood to the right of the bridge; the other, to the left.

And Leah had to choose. To find Boris, she would need to go to one of the houses and ask for help. But which one?

She crossed the bridge and rapped on the door of the house on the right. Almost immediately it was opened.

"Come inside. Quick." The frightened man pulled her across the threshold and slammed the door.

"Who told you to come here?" he asked. "Why didn't you go to the other house?" he asked.

"I just had a feeling," she explained. "In my heart." Leah knew it sounded absurd, but it was the truth. She had no further explanation for why she chose one house and not the other.

"Well," the man said, "your heart saved you." In the house on the left lived a family of Ukrainian policemen.

He took Leah out the back and to the barn. When they were inside, he turned and closed the door. Leah was frightened. He told her to climb up the ladder to the hayloft. She hesitated. "Hurry," he said.

Climbing the ladder, Leah heard a rustling in the hay. Someone was up there.

The man stood behind her by the barn door. She did not want to go up to the loft. And she did not want to go down, either.

Leah felt like crying, but no tears came. She continued up the ladder.

In the loft, behind the bales of hay, was another girl. It was a schoolmate from Reflovka, Feigale. She had been smuggled out of the ghetto before the soldiers took everyone away. She had been hiding here for a week.

The girls looked at one another and hugged and cried and laughed.

Leah was stunned, yet filled with an unexpected joy. She had followed her heart, knocked on a door, and entered into the past. This is what it must be like to go to Heaven and be reunited with those you had loved.

She was no longer alone.

* * *

That night when the man brought the girls their dinner, he said, "I can't keep both of you here. My wife wouldn't allow it." One girl could stay, the other must go. The choice was theirs.

Leah considered her predicament. Another reassuring world would be broken apart. Repetition did not make it easier.

"Do you know where Boris Savchok lives?" she asked.

The man returned later that night. He led the way, and then he left. Once again, Leah stood in front of a strange door.

She tried to put the reunion with Feigale out of her mind. She should have known better than to expect protection and friendship. Yet if Boris turned her away, it would be a disappointment worse than anything she had ever known. It would mean she had come to the end of all her possible inventions. There would be no place to go; and she would be doomed.

She knocked on the door, and waited in the darkness.

TWENTY

Just days before the Passover holiday, Peltz's company was on the front line eating their tasteless evening meal when a runner arrived from Command. Peltz read the communiqué and ordered a sergeant to get a patrol together. They were to cross the Senio to a building the Germans had been using as an artillery observation post. "Intel thinks the place has been evacuated. But they want to know for sure. Check it out and get back to me," Peltz said. "I'll be waiting for your report."

A dozen men were chosen for the patrol, including Arieh Shechter and Joseph Schneur. They were best friends. They had volunteered for the Brigade because they considered it their duty, as Jews and men, to fight the Nazis. But they were not by instinct soldiers. Even in the field, Shechter carried a duffel bag loaded with books. After dinner the two friends often would talk for hours, sharing ideas. But that evening they were armed, and wading through the cool, shallow water of the Senio.

Once across the river, the patrol moved cautiously in the twilight. The observation post, formerly the home of a wealthy family, had been built perhaps a century ago from pale yellow blocks of irregular stones. It cascaded down the hillside in a progression of broad formal terraces, and the views across the Senio were magnificent.

A finely pebbled path led to the entrance portico, and the soldiers approached in silence. It was easy to imagine the grandeur and elegance the house would have conveyed to a visitor before the war. But now dark and shuttered, ominously quiet, it provoked more disturbing thoughts: inside was a warren of rooms, a maze of shadows and hiding places.

The patrol entered through a pair of lopsided wooden doors that were hanging wide open on their hinges. It was a welcome that offered two very different interpretations. Either the Germans had fled in a careless hurry. Or, it was a lulling invitation into a trap.

In the large, dim hallway the men quickly separated into three squads. One went up a curving central staircase. Another, led by Shechter, followed a corridor to the right. Schneur guided the remaining squad into a passageway on the left.

Their inspection was slow and deliberate. The enemy could be waiting anywhere. The knob on a door would be twisted, a room would be entered, and suddenly a closet would fly open, or a curtain would lift—and the Germans would open fire. The men were charged and alert. Fingers poised on triggers, they moved ahead.

Shechter led his squad around a corner and into a dark corridor. Waiting for them was Schneur's squad.

Confused and frightened, the men began shooting. In the confined space the report of the bullets was explosive. By the time the frantic shouts of "Cease fire! Cease fire!" could be heard above the noise, Shechter and Schneur lay bloody and motionless on the floor. The two friends had been killed.

The patrol found no Germans in the house. Schneur and Shechter were the only casualties. Their bodies were strapped side-by-side on the hood of a jeep for the trip back to the field morgue.

Peltz reported the deaths to Command, but his account did not begin to reveal how deeply he was affected. He could not push what had happened from his mind. He wanted to make sense of it. He needed to assign blame.

He told himself that although he had sent out the patrol, the tragedy was not his fault. Schneur and Shechter should not have been soldiers. They were not men like him.

He had been imprinted by a different mold. From childhood he had been taught to ride, to shoot. His family owned land, thousands of acres. He was the grandson of a circus strongman who could bend iron bars in his thick hands and the son of a soldier decorated by the kaiser. He had little spiritual life; he was a Polish patriot, not a Zionist.

And who were they? Shechter and Schneur were, Peltz imagined, the sons of ghetto Jews. He knew the type. Bookish, detached intellectuals, raised in the tradition of medieval religiosity to be meek and passive.

No, he was not like them at all.

Three days later it was the first night of Passover, the holiday commemorating the exodus of the Jews from Egypt. Over a long, rigidly orchestrated meal, the seder, the story was retold of the Jewish people's escape from Pharaoh's oppression and their journey to a Promised Land.

It was an ancient biblical story filled with great drama: cruel plagues descending from the sky, a surging sea parting and then swallowing an enemy army, food falling from the heavens. Yet its core was a political narrative about the transformation of a tribe of slaves into a nation, a people with laws and a homeland.

"In every generation the Jew must look upon himself as if he personally had gone out of Egypt," the service instructed. When the Brigade sat down under the last pale light of a blue Italian sky, as dozens of seders were conducted along the front in a war against an enemy more evil than any pharaoh, the lesson in these words was inescapable. The attachment to their ancestors was direct and powerful.

As the sun faded and services began, the German guns opened

fire. At first the shelling was sporadic, and the men paid it little mind. The nights on the front were often punctuated by small disturbances. But soon the hammering of the guns grew more ominous. The scattered enemy guns had joined together in a massive coordinated attack on the Brigade's exposed eastern sector. Cannon and mortar shells exploded with continuous noise.

The soldiers under fire tried to continue their seders. They raised their voices in song and prayer as though hoping this act of common will could drown out the noise of the enemy guns.

But when the shelling did not stop and the echo of the explosions reached closer and closer, they took cover. Like their ancestors in the wilderness whom the Bible said began "murmuring" after all the relentless hardships, many of the men grew apprehensive. And they began to wonder what more might be demanded of them before they, too, would be allowed to journey to the Promised Land.

On the front line, it was the cruelest night of the war, a time of maimings and deaths.

Peltz was observing Passover with his company in a small farmhouse perched on a grassy humpbacked hill about four or five hundred yards from the Wehrmacht positions. Two days ago German artillery spotters had peered out from the windows in the dusty front room. This evening the room was crowded with Jewish soldiers.

Narrow wooden planks were placed on the floor and blankets were spread over them as tablecloths. Holiday candles were balanced on helmets. The men, like the ancients, sat on the floor. From somewhere—"manna from Heaven" one of the men quipped—a bottle of wine was produced. The blessing of thanks for the fruit of the vine was offered, and the seder began.

As the first of the Four Questions was about to be asked, the German guns started to fire. But unlike elsewhere on the front, in this farmhouse the noise was distant and unthreatening. "Why is this night different . . ." a soldier began to chant.

Peltz, thoughts about Shechter and Schneur's deaths still running through his mind, found he was caught up in the service. It all affected him: holiday candles flickering in a small hilltop farmhouse; the chanting of solemn age-old prayers; the steady pulsing tattoo of artillery; men with the Star of David on their uniforms honoring their ancestors who had defied a pharaoh and his army.

Swept along by these sounds and images and devotions, he began to question his previous certainty. Was he really that different? Another sort of man? Dismissing Shechter and Schneur's deaths as an isolated accident, a tragedy shaped by a single impetuous pull of the trigger, was too simplistic. There were more powerful forces at work. From the moment of their births as Jews, a determining chain of events had been set in motion. It was inherent in the binding terms of their covenant. And it was his inescapable and defining birthright, too. He could neither hide from nor deny it. That was the overriding lesson of this war. Of these prayers. Of the two friends' deaths. He was another Jew walking down a long dark corridor toward a communal destiny.

Carmi's company was spread out in the lowland trenches along the Senio, and on Passover afternoon the supplies—prayer books, matzohs, the promised extravagance of wine—still had not arrived. Command told them not to be concerned. A convoy of mules was on its way. The animals would be loaded with both ammunition and packages for the holiday.

It was nearly sunset before they saw the long line of mules, crates tied to their backs, being slowly led by a pathfinder down the rocky slope.

"Soon as we unload the mules, we'll say kiddush," Carmi told the men. They had spread a blanket over the bottom of the trench for their Passover table.

Moments later the German guns began firing. Explosions pounded the hillside as the mule train descended. Shells tore into the rocks. The mules brayed and bucked, pulling in all directions. The pathfinder had to struggle to keep a strong hand on the lead animal.

But as the noise from the shelling increased, it became more difficult to control the animals. The mules were kicking wildly, whining as if in agony. Then suddenly the mules broke loose. Crates tied to their backs, they raced down the mountainside toward the river.

Soldiers shouted from the trenches: "Get them. Don't let them get away."

Despite the artillery fire, men leapt from their concealed positions and ran after the galloping mules. But the animals were surprisingly fast. It was only when there was a pause in the shelling and a mule would come to a halt that the men could manage to catch one. But then there would be a new explosion, and the mule would kick out frantically and run off again.

By the time the exhausted soldiers managed to lead some of the mules back to the trenches, it was night. And after they started unloading the recovered crates, they discovered that one position had the matzohs, another had the prayer books, another the wine.

In the darkness, the shells still falling, the men ran between the trenches trying to exchange what they needed to make a seder. But it was futile. Too many mules had run off. There was not enough to go around.

When Carmi stood up in his trench along with his men and finally raised a tin cup holding only a small sip of wine for the kiddush, it was closer to midnight then sunset. He was annoyed. Not only was this night different from all others, but also this seder was different, too. Nothing was right.

As the service continued, he found his mind wandering to thoughts of Tonka, and the little one. He wondered how soon it would be before his daughter Shlomit was old enough to ask the Four Questions. And with these thoughts of home, he began to realize that this makeshift seder in a trench crowded with Jewish soldiers was making all the future seders possible. They were fighting so that one day they could go home and sit around tables with their families in peace. With a newfound appreciation for this untraditional gathering in the darkness near the banks of the Senio, Carmi raised his voice in prayer and thanks to the heavens.

* * *

Pinchuk attended the seder at the battalion headquarters. It was outdoors, under a warm Italian sky sprinkled with stars. The large table was set with bottles of wine, tall stacks of matzohs, and large platters of roasted meat. At the end of the meal the traditional holiday desserts of macaroons and kneidlach were served. The front was so far off that the German guns were just a faint, distant rumble.

Yet Pinchuk could not find it in himself to enjoy the holiday dinner. His mind was full of the seders he had celebrated in Reflovka. At the time, those evenings had been unappreciated; another beat in the rhythm of his life. But now when the image of his family seated at the Passover table rose up in his mind, he was nearly overcome. Looking back, he saw that it was a period of perfect contentment. Only it had been taken from him. And unless he could somehow find a way to retrieve what had been stolen, to re-create the harmony of those moments, Pinchuk was certain he would never know peace again.

TWENTY-ONE

The door opened and Boris stared at Leah. He did not say anything, and for a terrible moment Leah lost all hope. But then Boris reached out and took her into his arms. She cried, and he cried too.

She lived in Boris's barn. The days were hard; there was no one to talk to. Nights were even harder. She lay curled up on a thin straw bed, trying to push her thoughts about the past away. Memories were a strand of hope; and hope was irrelevant.

This was her life, and how she survived.

On the tenth day, Boris came and told her that she would have to leave right away. The police had heard rumors that Jews were being hidden in the village and were making a house-to-house search. Soon they would be here.

"I'm sorry," he said.

"I'd better hurry," Leah said finally.

It was colder in the forest than Leah remembered. The next day Boris brought food and a new plan. A group of Jews who had escaped the roundups would occasionally come to his house in the evening to beg. Yesterday when they knocked on his door, he had asked if they would allow Leah to join their group.

They are waiting for you, he said.

Only one man was standing by the grove of birch trees. He was short, not much taller than Leah, and his clothes were covered with dirt. He did not introduce himself but Leah recognized him as the *shohet*, the man who slaughtered the kosher meat, in the town of Olizerka. Whenever Leah looked at him, he looked quickly away.

Boris handed him a few coins, then gave Leah a farewell hug. He was forbidden to come any further. The Jew's hiding place was a secret.

The mist was a thick mask over Leah's eyes. She had followed the man for hours, and now they came to a swamp. It looked wild, as though it could swallow up anyone who fell in without leaving a trace. Leah did not see how they possibly could go any further. But the shohet told her to hold on to his coat, and then led her along a hidden path of wooden planks placed end to end. Leah clung to his coat until they were across.

Beyond the swamp, the land was dense with high pine trees. It sloped downward, and soon Leah began to smell smoke from a fire.

She followed the shohet into a clearing. Seated around a glowing campfire were about twenty people.

They became silent as she walked toward the fire. In the amber light, she stood before the group. No one spoke to her. Leah was not sure what she had expected, but she was unprepared for this reaction from a group of Jews. She realized at once that she was not welcome.

What is the difference between Heaven and Hell? her brother Arie had once playfully challenged. In Hell, the rabbis say, we're seated at a long table filled with delicious food. But our arms are chained. The chains are loose enough so that we can reach out for the food, but tight enough so that we cannot bring the food to our mouths.

In Heaven, it is the same situation. The same banquet table. And the same constricting chains. But in Heaven, Arie had explained, people had the spirit to realize they could raise their arms high enough to feed the person next to them. People helped one another, and it was a perpetual feast.

Leah found herself remembering this parable because, as time passed, she began to fear she had been consigned to Hell. Burdened by the weight of their own heavy chains, no one in the group helped her. She was shackled, and she was on her own.

When the weather started to change, Leah feared she would not survive the winter. Her feet were wrapped in paper. She did not have a coat. What would she do when it snowed?

During the long empty days, it was always on her mind. At night, she dreamed of a frozen corpse lying in the snow.

Leah made up her mind to leave the camp. She would go to Boris and he would give her a new pair of shoes. Sturdy leather shoes with laces. Perhaps even a warm coat, too. When she was outfitted for winter, she would return.

Except Leah knew she could never find the way to his house. She would get lost in the forest.

There was a brother and sister from Olizerka in the group. The girl's name was Sarah and she was just a little older than Leah. She would be able to find the trail into the town. Leah begged Sarah to come with her.

She refused.

Leah pleaded and did not stop until Sarah gave in.

They were past the swamp, but still deep in the woods, when it started to snow. It quickly developed into a powerful storm. Leah slapped her hands together to try to keep warm. The snow was already seeping through the wet paper wrapped around her feet. Her soles were growing numb.

The force of the storm filled Leah with awe. With all this war demanded, how could there still be enough power in the universe to unleash such ferocity?

Sarah became lost. The snow was too disorienting. So Leah led the way. Sarah walked in the footprints that Leah made in the fresh snow, and Leah could only guess where she was going.

Leah would not allow herself to stop. But each new chill made her tremble. Her dream had been a prophecy. She would freeze to death.

It was after midnight when the two exhausted girls reached Olizerka. It was still snowing. In this brutal weather Boris's home might as well have been in another country. They never considered going any farther.

They spent the night in an abandoned house. There were many in the town since the Jews had been taken away.

In the morning, when Leah went outside the snow reached up to her waist. She was hungry and she needed a pair of shoes.

Sarah told her that in the past the group had gotten food from a family in the town who were members of the Shtunditz sect. Their religion made them sympathetic to the Jews. They could go to the house and beg for help.

The family offered them food, and gave Leah a new pair of bark sandals. They were not the lace shoes she had imagined Boris giving her, but she was satisfied.

"Now we can go back," said Sarah.

"No," said Leah.

In her mind the picture of the blue frozen corpse had risen up again. She told Sarah that she wanted to spend the night in the abandoned house.

"Someone will notice us," Sarah argued.

Leah would not give in. They would wait one more day. By then the snow will have begun to melt.

* * *

They left at dawn.

As soon as they crossed the swamp, Leah sensed that something was not right. When she did not see the fire, she grew more anxious. She walked hesitantly into the camp. Her footprints marked her slow path across the snow. Sarah followed.

"Leah!" the girl screamed. "You're stepping on someone."

Leah looked down. She was standing on an arm. She was trembling, but she bent down and forced herself to brush away the snow. She saw a face. It was Sarah's brother. He had been shot.

Sarah screamed, and the agonized cry pierced the forest. But with surprising calm, Leah continued to hunt through the snow. Lying nearby were five more bodies. All had been shot. The skin of each had turned a milky ethereal blue. The color of the frozen corpse in her dream.

Sarah sobbed all the way back to Olizerka. They were once more on their own. Two girls without protection, without prospects. But Leah felt her instincts had again saved her. Somehow she would survive.

TWENTY-TWO

The fateful Passover week—five dead and fifty-two wounded—left Carmi eager for the final push across the Senio. He wanted to extract revenge for these casualties. But on the third of April the battle plans for the spring offensive were still not finalized, and he was summoned back from the front and ordered to report to Command. Moshe Sharret, the head of the Political Department of the Jewish Agency, had come from Palestine to present the Brigade flag to Brigadier Benjamin. Each unit had its own delegation at this event, and when Carmi arrived he was glad to see that Peltz and Pinchuk had also been selected. He hurried across the parade ground to join them.

It was a fresh spring afternoon. The ceremony took place in a grassy meadow behind the ramshackle villa that served as headquarters. Enemy artillery thundered sporadically in the distance. Sharret and the brigadier stood side by side facing the orderly rows of men from behind a wooden table that had been hammered together earlier that day.

Sharret was to speak first, and he stepped in front of the table to talk more directly to the troops. He was an unprepossessing man, short; a drab clipped mustache grew beneath a prominent nose. In his baggy wool suit, V-neck sweater and tie, a fountain pen sticking from the top of his breast pocket, he seemed out of place in the company of soldiers. He look tired, a bureaucrat more accustomed to spending his days in airless rooms poring over the tiny print in obscure documents.

But when he delivered his address, it was as if he had been transformed. His hands were clenched into fists and he held them rigidly at his side like an officer on parade. His voice bellowed out across the countryside, alive with resolve, commanding:

"At last Jews can see the flag for which they have always fought in their hearts. It represents the Homeland and also the blood of those who in their millions have died without having the chance to fight back. . . ."

Carmi listened and found himself thinking back to the last time Sharret had addressed the troops. It was in the Haifa cinema before they had left for Egypt. That assembly had ended in anger after the British colonel stomped out during the singing of "Hatikvah." He remembered, too, how it had taken a mutiny to raise the blue-and-white flag above the barracks in Benghazi. And now the flag with the Star of David at its center would be carried by Jewish soldiers into the final battles against the enemy. What once had been an act of defiance had become institutionalized. So much had been accomplished in the past two years.

When the Second Battalion had been sent to the port city of Benghazi in North Africa in July 1943, the troops had believed that the Palestine Regiment would soon be embarking for Europe. However, months passed and the monotonous routines of parades and guard duty continued. And it became apparent that the British strategy was simply to get the Jewish troops out of Palestine—not closer to the war.

The mood in the camp, then, was already uneasy and resentful when the British battalion commander, Col. Best, making his inspection of the barracks, noticed the Yishuv's national flag affixed to a wall. He ordered that it be taken down. When the soldiers hesitated, the colonel ripped the flag from the wall.

To the colonel, the flag was an expression of the Zionists' hope for statehood, and an insult to the Empire.

To the men of the Brigade, though, it was more than a political banner. At a time when Jews were being sent to their deaths in Europe, it was a defiant proclamation: The people of Israel live.

When the troops heard that the highest-ranking British officer in the camp had torn their flag from the wall, the battalion's Haganah leadership called a meeting. They talked into the night, but they could not agree on any action.

The next morning after reveille the men assembled as usual in the camp's parade ground, a dusty plain surrounded by the four troop barracks. Each day they would salute as the Union Jack was raised to the top of a long, white pole on the roof of the west-facing barracks. Today when the men fell into ranks, they discovered that a flag was already flying. It was blue-and-white and in its center was the Star of David. And Israel Carmi had raised it.

When the colonel ordered Carmi, as the battalion sergeant, to accompany him to the barracks roof, they found the flagpole surrounded by a wall of men. The colonel was unprepared for this development, but it was no surprise to Carmi. He had assigned the platoon to protect the flag.

"Sergeant," Carmi heard the colonel bark, "I order you to lower the flag."

The soldiers were bunched in front of Carmi as though they were in a rugby scrum. He placed his huge, heavy hands against the shoulder of one of the men and pushed. There was more irony than muscle in the shove, yet Carmi, when the man did not budge, reacted as if he were astonished by its ineffectualness.

"Sir," he said, "your order cannot be executed."

Col. Best stared at the men. He looked up at the flag. At last he turned smartly and walked back to the door.

By the time the colonel had returned to the parade ground, the blue-and-white national flag was flying from nearly every pole in the camp.

Before the morning was over, the British had declared the situation a mutiny. After consulting with his superiors, Col. Best announced that the entire battalion would be stripped of its weapons and placed under arrest if the Zionist flags did not come down.

Not a single blue-and-white banner was lowered.

It was a tense standoff. Both sides were armed. The first shot, fired in anger or premeditation, by either a British officer or a Jewish soldier, would lead to a disaster.

As the anxious day stretched on, the Haganah commanders began to acknowledge among themselves that their intransigence, however valiant, would be self-defeating if it kept the battalion out of the war. They decided to put the matter to a vote. The entire battalion met in the hall where a day earlier the catalyzing incident had occurred. The vote was nearly unanimous. The flag would come down, but it would be raised again on holidays.

That evening the battalion lined up in threes on the parade ground. Col. Best and his British officers watched the assembly from the roof of an adjacent building.

After ordering the men to attention, Carmi went to pass off the parade to the most senior Hebrew officer present. This was the usual procedure, but the officer declined. "It's your parade, Sergeant," Capt. Aharm Hoter-Yishai announced.

Carmi ordered the bugler to play. As the men stood at attention, Carmi and the officers saluted the blue-and-white flag.

The banner was lowered and Carmi folded it with precision.

"You may dismiss the troops, Sergeant," the captain ordered.

But Carmi hesitated. From the rooftop, the British officers watched, wondering what would happen next. The men waited, too.

Carmi began singing:

"Kol od baleivav p'nima . . ."

His voice was loud and heartfelt. The entire battalion, over a thousand men, joined in. They stood at attention on the parade ground and sang "Hatikvah," The Hope. It was the Yishuv's national song.

"*. . . Nefesh Y'hudi homiya . . .*"

When the anthem was completed, Sgt. Carmi announced, "Troops dismissed." The mutiny was over.

Now standing at attention in a field in Italy, Carmi vowed that the day would come when their flag would fly over an independent Jewish state. Days ago, in a trench by the Senio, he had heard the story of the exodus, the birth of the Jewish nation. The past, he knew, carried with it a plan for the future.

After his address, Sharret presented the carefully folded flag to Brigadier Benjamin. The troops snapped to attention as the oldest enlisted man in the Brigade unfurled the banner. A bugle sounded, and the flag was raised up the pole.

With the blue-and-white banner fluttering against the Italian sky, the troops sang "Hatikvah." Some of the soldiers wept openly. Carmi, though, was moved only to commitment.

When the troops dispersed, Carmi spoke to his friends. "Well, Johanan? I *heard* you. You were singing 'Hatikvah.' Thought you believed Jews had to behave like good British soldiers."

Carmi shot Pinchuk a mischievous wink. It was a signal meant to say, "Let's have some fun at Peltz's expense."

Pinchuk knew Carmi was just teasing, but he felt uncomfortable mocking a man who had led a bayonet charge against the Nazis. Not when all Arie Pinchuk had done in the war was to make sure the trucks were running. He had sung "Hatikvah," but he had not fought.

Pinchuk excused himself. There was, he said unconvincingly, a situation back at the motor pool that needed his attention. With a perfunctory wave, he hurried off.

Pinchuk's abrupt departure left Carmi feeling disturbed. He liked Pinchuk, but there was something unsettling in the young officer's remoteness. At first he had thought it was simply a natural shyness. But there was also a morose quality to Pinchuk's detachment. And Carmi could only wonder at its cause.

But Peltz drew him out of these thoughts. Peltz was not shy. And he was not a man to be teased. Carmi noticed that his friend had pulled himself up to his full height. Perhaps it was an unconscious rebuke, but Peltz, lithe and saber-straight in his pressed officer's dress, now seemed to tower over the squat, thick-muscled sergeant.

"Singing 'Hatikvah'? You must be mistaken, Israel," he lied.

Later, as Carmi was leaving to return to the line, he felt someone tap his shoulder. He turned and saw a lanky, blond, blue-eyed soldier. It took him a moment to recognize the grinning face. Then he realized why. The last time he had seen Oly Givon the man had been wearing the uniform of a Wehrmacht SS captain.

That had been three years ago, outside a cave in the woods near Mishmar Ha'emek. It had been a time when it seemed certain that Rommel would soon march into Cairo, and then on to Palestine. As part of its desperation strategy, the Yishuv had assembled a platoon of German-speaking Jews and trained them in great secrecy to impersonate the enemy. When the German troops goose-stepped down the streets of Tel Aviv, these impostors would be sent out on bold, last-gasp sabotage missions. Carmi, already a sergeant in His Majesty's army, had been chosen as the liaison between these Jewish commandos and the regular British forces.

As the war had played out, Rommel was pushed back at El Alamein and the secret German Platoon, as it was known, was never called into action. But Carmi could still recall how disturbing he had found their perfect imitation of Nazi soldiers. It was not simply the bark in their guttural speech or the threat in the swastikas on their uniforms. The men had even moved with a heavy, arrogant swagger that, in Carmi's susceptible mind, provoked visions of the Gestapo dragging their passive victims away. Carmi boiled, and he needed to remind himself continually that these troops in their grim battle-gray uniforms were in fact Jews. It had been an unpleasant assignment.

"Oly, *shalom, shalom,*" he said as he shook the man's hand. "What are you doing here?"

Givon explained. The British War Office had asked the Yishuv for reinforcements to fill out the Brigade before the final campaign. When Ben-Gurion and his advisers met to consider this request, it occurred to them that it might prove valuable to have members of the German Platoon in Europe.

"So," said Givon. "Here I am." He had a cheery way of talking that was expansive, yet also guarded. Carmi, whose own life in the Haganah was an arsenal of secrets, did not press.

"L' hitraot." See you. "And stay safe," Carmi said. They shook hands and Carmi left. By the time he returned to the trenches, this chance, brief meeting was long out of his mind.

TWENTY-THREE

Exactly one week later the Brigade attacked.

WAR DIARY
Unit: *HQ JEWISH INF BDE GP*

April 10. EIGHTH ARMY general offensive commenced. 2 Pal R crossed R SENIO seized and occupied objective FANTAGUZZI M213231 by 0145 hrs. No casualties.

The breakthrough was designed to be forceful and direct, and the first stage was executed with precision. Just before dawn, Carmi watched with awe from his trench as a thousand American bombers rushed overhead. They came in ten waves, one hundred planes to a wave, and when their bombs started falling in vicious syncopated succession on the enemy positions, the ground in his distant trench rumbled. "They are plowing the land inch by inch," one of his men yelled.

After the planes had finished, the big guns let loose a brief, sustained burst of fire. Then all at once the guns stopped.

It was a pause of unnatural quiet. A blank space. And suddenly the order was shouted up and down the line: "Forward!"

The men rose up screaming from the trenches. The Brigade engineers had set up a series of pontoon bridges across the shallow river and the troops charged across. Carmi's company was in the lead, and as they climbed up the north banks of the Senio, hearts pounding, they were prepared for fierce resistance. But there was only a drizzle of enemy machine gun bullets, and these pillboxes were quickly pounced on. It did not take Carmi long to realize the Germans had withdrawn. The Brigade had taken the north side of Senio.

The bridgehead beyond the Senio was vital. It was the high ground that would allow the Fifth and Eighth armies to converge swiftly on Bologna, and then move on to trap the enemy against the Po.

On the Brigade's right flank were the Polish Corps. They had attacked simultaneously with the Jewish troops and had also succeeded in chasing off the Germans. The Italian Folgore Regiment was on the left flank. Their attack that morning had failed. The Germans had kept their hold on a high, mountainous sector just beyond the river.

It had to be captured. The Allied Command feared the Wehrmacht would be able to use this strategic position, particularly the artillery on the crest of Mount Ghebbio, to challenge the entire bridgehead.

If the Germans regained control of the Senio, the carefully designed tactical hinge would snap, the Allied armies would be diverted, and the enemy would escape intact across the Po. It was decided that the feckless Italian regiment would be reassigned. Another more spirited unit would have to take out the guns on Mount Ghebbio.

WAR DIARY

April 12. JEWISH BDE given task by 10 CORPS of occupying M GHEBBIO 2023.

Peltz stood on the narrow dirt road and looked up at Mount Ghebbio. It was more a steep green hill than a mountain. There were many mulberry trees, and fields dense with tall shoots of wheat. On the summit was an ancient stone church. The morning sun flashed on the white cross high on its roof. Peltz could imagine setting out with a pretty dark-eyed Italian farm girl for a hike up to the top on a warm new spring afternoon like this. But today he was going into battle.

The morning had been uneventful. The troops from the Third Battalion had quickly crossed the river and moved past the positions where Carmi's Second Battalion had dug in two days earlier without meeting any resistance. They continued along an open road to the tree-lined base of the mountain without hearing a shot. Up and down the line, the men wondered if the Germans had already retreated.

Maj. Maxim Kahan would command the assault. Peltz had known him in Haifa and, quick to measure a superior officer against himself, he had bristled when he first saw the new major stride into the battalion mess. Then on a reconnaissance mission Peltz had watched with surprise, and admiration, as Kahan charged straight into a German pillbox. After that display, he only hoped that when the time came, he could demonstrate to Kahan that he possessed a similar courage under fire.

As they waited at the base of the mountain, Kahan turned to Peltz. "It's pretty quiet."

"Maybe they left. Heard we were coming."

"Only one way to find out. Ready?"

Peltz nodded, and Kahan gave the order to advance.

They went up the hill in a box formation: a lead platoon, the company commander and his staff, and then two more platoons, one directly behind the other. Peltz was in the forward platoon, and he took each new step waiting for the enemy to open fire. But there was only the steady thrashing sound of the troops through the high grass and wheat. The men did not dare to speak. It was as if they believed a single spoken word would undermine the incredible possibility that the Germans had abandoned their positions.

Suddenly there was an explosion. And another. Peltz recognized the nasty thuds of Schu mines. Then he heard the agonized screams of a wounded man. And he realized with horror that the hill was mined.

His platoon froze in place, anxiously examining the ground while a medic and a stretcher bearer ran toward the howling cry. Minutes later they returned carrying a private on the stretcher. The man's foot had been blown off. His shin ended in a jagged, bloody tangle of flesh and bone.

The men parted to let the bleeding man down the hill, and he passed through the troops as if on parade. As he went by, the wounded soldier looked up from the stretcher at his friends and seemed to gather strength. He spoke a single word, but he repeated it like an incantation. "Revenge! Revenge! Revenge!"

Without pausing to consider the risks, Peltz ran to the front of the lead platoon. With his rifle held high, he yelled, "After me!"

He charged forward, up the hill, and into the heart of the minefield. Screaming with determination, roaring and fearful, the men followed.

The enemy opened fire. They had not retreated. They had been waiting, setting their trap.

The Brigade had entered into the heart of the German defense. But they did not retreat. Defiant, they tried to burst through it.

There were snipers and mortars, and the Schu mines were exploding, and yet the men kept climbing. "After me!" Peltz repeated wildly.

The ascent was a climb into a swarm of bullets, the air laced with flying metal. But Peltz was too excited to stop. With his long strides, he raced through the minefield, toward the Germans.

About three hundred meters below the church, the hill banked, and both Peltz and Kahan dove into this crease in the terrain. Kahan ordered the signalman crouching beside them to call headquarters. Kahan gave the artillery command their position and asked for close support. "Pound those bastards," he yelled into the phone.

The signalman's head was raised above the ditch, his radio antennae sticking high up into the air. "Danny," Peltz said, "get your head down! You're not going to do us any good with a bullet in you." As soon as he spoke, Peltz heard a sharp *ping!*, and the antennae tilted toward the ground at a precarious angle. The signalman's inert body hit the dirt heavily. He had been shot dead.

The battle continued. All Kahan could do was order another man to unstrap the radio from the dead man's back.

The Brigade artillery, forty-millimeters usually used only as anti-aircraft guns, began hitting the German positions. The men huddled along the dip in the terrain as the shells screamed overhead. The sound of explosions filled the troops with confidence.

Kahan quickly divided his force into two attack groups. They would charge forward with bayonets. One would flank right, the other left. After they had gone two hundred meters, the remaining troops would rush up the middle.

"And when we get to the top, Maxie?" Peltz asked.

"We push the bastards back to Germany," Kahan said.

Peltz tried to rein in all his anxiety as he rammed the long bayonet onto the barrel of his rifle. The sound of the blade clicking into place helped him focus.

Moments later, bayonet held high, he leapt from the ditch and charged toward the church. He was screaming incomprehensible sounds from unfamiliar depths of his being. He was beyond fear. It was an emotion that in that long mad rush up the hill had no meaning. It would have been too human, too cognitive. He could not stop because that would have required some control. He rushed on.

At the top of the hill, a German ran toward him. The two men closed on each other like jousting knights. Peltz went into a crouch. The German swung his rifle butt at Peltz's head. Peltz lunged, stabbed the soldier in the stomach. The man went down. Peltz pulled his blade out, and joined the others as they swarmed into the church.

Peltz ran in ready to fire, but the Germans had retreated. The enemy artillery had been abandoned, and the Allied troops could advance. The Jews had taken Mount Ghebbio. The Brigade had driven the Germans off.

* * *

The rout continued throughout Italy. The Allies moved through Forlì, Imola, Castel San Pietro. After some hard, costly fighting, they took Bologna. The German strategy of retreating to its fortress beyond the Venetian Line for a last stand was never realized. With the escape routes into the Alpine passes blocked, frightened, weary German soldiers began surrendering to the advancing Allied troops.

On April 14, however, the Brigade was ordered to halt on the outskirts of Bologna. The Allied armies proceeded up Route 9, past blazing carpets of poppies, toward the retreating enemy, without the Jews.

The men of the Brigade had been filled with anticipation. They had been prepared "to push the bastards back to Germany." They had endured the heat of battle against their blood enemies and wanted more. Yet just as they imagined extracting victory and a measure of vengeance, they were stopped. And it left them unsatisfied and tormented. The suddenness of the order that pushed them to the sidelines was an insult to their pride.

British policy toward Palestine, not military strategy, had determined this decision. In its brief time at the front, the Brigade had been bloodied. There were 57 deaths, and 150 wounded. His Majesty's government decided that its tenuous relationship with the Jewish Agency would come apart completely if it appeared that Jewish soldiers, young men who would shape the future of the Palestine territory, were being deliberately sent off to die.

The Brigade had been out of the hostilities for nearly three weeks, glum and frustrated spectators watching as the war sputtered to an end without them, when the German surrender in Italy became effective at six P.M. on May 2.

"I won't go," Carmi vowed.

He was lying on the grass under the shade of an olive tree. In the distance, shimmering in the May morning sun, was the shiny blue Lamone River. Peltz sat across from him, his back squared against the tree. It was two days after the war had ended in Europe.

"I'll desert," Carmi told his friend.

Peltz also felt angry and betrayed. He was still raging over the War Office's refusal to let the Brigade fight in the final battles of the war. He had been prepared to charge, his bayonet held high, all the way to Berlin. He was a Jew and there were still scores to settle. And now a rumor was spreading through the Brigade that they would be sent to Burma.

Peltz had not volunteered to fight against Japan. That was not his war. He would desert, too.

Carmi understood what the manipulative British government was trying to do. They were determined to keep well-trained Jewish troops out of Palestine for as long as possible. His duty, however, was to get back home and prepare for the battles that would come.

But Peltz had another reason for refusing to go to the Far East. His mission in Europe was unfinished. He had to get to Zabiec. In his mind he had tried to hold the idealized picture of his riding up to the grand estate, the paradise of his childhood, in his British officer's uniform and embracing his mother and grandfather. But it was becoming increasingly difficult to sustain such a hope. When the men had visited Bologna they had returned to camp with the report that it was impossible to find more than a handful of Jews. The rest had been rounded up, taken away. And that was in Italy. It was painful to contemplate what might have happened in Poland.

Waiting these past weeks in the hills of northern Italy, Peltz had begun to appreciate that he was on the edge of knowing what had become of his family. His excitement had grown into an almost physical yearning. He could not allow them to send him to the Far East. Like Carmi, he would have to find his own solution.

But just then Pinchuk came running over. He looked more animated than either of the men could ever remember seeing him.

"Have you heard? Have you heard?" he asked excitedly. "The Brigade's going to Germany."

PART IV

TARVISIO

Summer 1945

TWENTY-FOUR

Deep in the Chernigov forests of the Ukraine, Leah Pinchuk stood guard. A rifle was slung over her shoulder and there was a green beret on her head. It was after midnight, the start of her four-hour shift, and she walked back and forth along a well-worn path.

"Use your ears, not your eyes," the tall Russian officer had instructed her nearly eight months ago. That had been her first night on guard duty and she had felt completely unprepared for the responsibility. But she no longer doubted her ability to do her job. She would fire a warning shot to alert the camp. Then, kneeling, her rifle held steady and pressed against her shoulder, she would wait at her post, prepared to shoot the first German soldier she saw coming through the trees.

The girl who had been afraid to sleep without a candle glowing in the room no longer existed. In her long ordeal Leah had found the will to do whatever was necessary. And now she was a partisan.

When Leah and a tearful Sarah had returned to Olizerka after discovering the frozen corpses in the woods, the Shtunditz peasants took pity and allowed them to stay in their barn. But two days later the girls learned the police were searching for them.

The peasants were terrified. They arranged a meeting that night—there was no time for delay, they insisted—with the partisans.

Sarah was afraid. "They're Russians," she said. "They'll do terrible things to you."

But Leah could not see another way. She left Sarah. It was another farewell in what was becoming a lifetime of good-byes.

"Can you kill Germans?" the Russian officer demanded.

"Yes," Leah answered. She had doubts, but she knew it would not be wise to provoke him.

It took two days to walk to the camp. In that entire journey, the Russian never said a word about her being Jewish. After all Leah had suffered, after all she had endured, Leah found this incredible.

In one world, her identity had been defined by an ancient covenant. Here there was a new one. If necessary, she told herself, she would kill Germans.

She had imagined a camp like the one in the woods beyond the swamp. But the partisan base was bigger than Reflovka. It was home to about three thousand fighters. There was an airfield, a hospital, a dining hall, a laundry, and, a luxury that seemed truly unbelievable, showers. She could wash herself with soap in her own stall. She looked around at this community—her new home—and felt the satisfied excitement of someone who had gambled, and had won.

The commander of the group was Maj. Gen. Aleksei Fyodorov. When the first waves of German troops had marched deep into the Soviet Union, Stalin had urged his people to fight back. In an emotional radio address, he called for the formation of "guerrilla units . . . to blow up bridges and roads, damage telephone and telegraph lines, set fire to forests, stores, transports. In the occupied regions, conditions must be made unbearable for the enemy and his accomplices. They must be hounded and annihilated at every step and all their measures frustrated." On orders from Nikita Khrushchev, first secretary of the Communist Party in the Ukraine,

Fyodorov had parachuted into northern Volhynia to lead the guerrilla war against the invading German army.

The Fyodorov Partisanka had been in the woods for over two years when Leah joined. They had "hounded and annihilated" the enemy. And the war in the East had begun to turn.

Leah was told to share a tent with a thick-waisted, unsmiling Russian woman. The woman took one long look at Leah and started poking her fingers into the girl's hair.

"Lice?" the woman asked.

"Yes," Leah admitted. They had been crawling over her body for so long that she had grown to accept their presence. Now she felt ashamed.

The woman led Leah to a tree stump and told her to sit. Without asking, she snipped off both of Leah's long blond braids. Leah did not say a word. When the braids were splayed out on the ground, the woman attacked Leah's head until only a thin bristle lay very close to the scalp. Leah had been very proud of her blond hair. Now it was another thing that had been taken from her.

The woman laughed. "You look like a chicken." She gave her a green beret and Leah pulled it tight, down to her ears.

The first time Leah fired the rifle, kneeling on one knee as the Russian had instructed, the impact knocked her down. "Again," the Russian ordered. She got back into position and took aim at the white circle that had been nailed to the tree. This time she did not fall down. "Again," the Russian ordered. On the third day, she hit the tree. "Again," the Russian ordered. On the fourth day, she hit the circle. "Again," the Russian ordered. On the fifth night, she was assigned to guard duty.

During the day, Leah worked in the kitchen. The cooks were Russians who had been flown into the camp. Her job was to clean the pots and peel the vegetables. She did not mind. If something was rotten and was going to be thrown away, she could eat it. After a few months, when she showered she noticed that her ribs were no longer protruding.

The cooks talked to her about their families, their wives, their children, and their homes. Leah told them that her mother and father had been killed. She did not tell them about her brother in Palestine or her plan to go live with him. That was her secret, her last defense, and she guarded it.

While they worked, the men often sang:

> At the edge of a forest a large tree stands.
> Underneath it lies a partisan.
> The wind blows his yellow curls.
> And beside him stands an old lady, his mother.
> She wipes a tear and kneels down to her son.
> But the officer pulls her up to her feet and
> says,
> "Don't cry, Don't cry.
> Your son died a hero."

Without trying Leah learned the words, and soon she sang along. Her sweet high-pitched voice had always given her father great pleasure. And now the cooks encouraged her, too.

One evening three captured German soldiers were led to the campfire. The youngest was not much older than she was. He had blond hair and blue eyes. He was trying not to cry. Another prisoner was so fat he was nearly bursting out of his uniform. He was pleading, "I was forced to join the army. I didn't want to. I have a wife. Two children." The third man looked as if he were in shock, as though he could not quite understand what was happening.

All three were shot. The body of the blond-haired boy was still convulsing as a Russian woman removed his ring and watch.

For the first time, Leah began to wonder if she could accept the terms of her new life.

* * *

The woman who shared Leah's tent had a lover. At home she had a husband and children, but in Fyodorov she had fallen in love with a Russian officer. "I won't leave him," she would tell Leah. "I wouldn't be able to live."

Leah would listen, but she would not respond. She knew too much about yearning. She did not want to tell the woman that you can learn to live without the people you love.

The woman advised Leah to find a boyfriend, too. Leah could not even begin to explain how impossible that was. Her heart was vacant. And the danger was unthinkable. If she dared to care about someone, it would only end in disaster. That was the recurring lesson in her life.

Nevertheless, Leah soon found someone. Feigale, the girl from Reflovka whom Leah had discovered in the hayloft before going on to Boris's, had also made her way to the partisans. Their reunion filled Leah with a cautious joy. At last Leah had a friend.

Feigale was four years older than Leah, and she was flattered by the attention a Russian officer paid to her. He had confided to Feigale that although he was a Christian, he had been born a Jew. He asked her to meet him one night by the lake, on the far shore away from camp. Feigale showered, put on a clean blouse, arranged her hair, and went off to meet the Russian officer. She never returned. The next day her naked body was found floating in the lake. But it was wartime, and death was commonplace. Only Leah seemed to notice Feigale's death, and even she tried not to dwell on it.

As the Wehrmacht troops began their disorderly retreat from the Ukraine, the partisans took advantage of the confusion and captured a high-ranking German officer. They brought him to the camp for interrogation, but none of the Russians spoke German. Leah was fluent in Yiddish, and she was summoned to translate as best she could.

She went into a dank room, and sat on a bench across from the German officer. The man's hands had been tied behind his back, and there was dried, caked blood beneath his nose. One eye was swollen to a thin slit. Yet he held himself together with an impressive orderliness, his back straight, his jaw clenched. When Leah sat down, their knees touched. He did not move away.

A Russian major prowled the small room. Edgy and aggressive, he spat out questions for Leah to ask. He wanted to know when a train would pass on a bridge, when a convoy of troops would be moving.

Leah translated, and the German answered in a clear, steady voice. Leah suspected he had made up his mind to cooperate.

When the Russian major finished his questions, he removed his pistol from his holster. He was a regular army officer and he had a new Nagant. He handed the gun to Leah.

"Shoot," he ordered.

Leah looked at the German. His silence seemed to acknowledge that there was nothing either of them could do. He did not appear scared, only weary. Leah's finger was on the trigger; a bit more pressure and it would all be over. It would only take an instant.

But she could not bring herself to do it.

"He killed your parents," the Russian major taunted. "Shoot him."

Leah raised the gun. "I can't," she said finally.

The Russian grabbed the gun, pressed it to the prisoner's temple, and blew the side of the man's head off.

A kaleidoscope of skull and brain matter and blood splattered Leah's face.

"Clean up," the major snapped. Then he walked off, leaving Leah to deal with the corpse, and her tears, and her squeamishness.

Leah waited for the punishment that was certain to come. She knew she was guilty. She had violated her agreement with the partisans: She had refused to kill. Her anxiety built until it nearly crushed her. She had been tested, and she had failed. She tried to imagine what her punishment would be, and each new day's wait was another torture.

TWENTY-FIVE

On the evening before the Brigade left for Germany, when the troops were about to be dismissed from parade, Sgt. Israel Carmi stepped forward. This was not part of the regular ceremony and as he walked to the white chalk square in front of the flagpole, the men waited to hear what he would have to say.

In normal times, Carmi's words would be listened to with attention and respect. It was well known that his pronouncements carried with them the authority of the Haganah. And, of no less importance to the young troops, they also bore the imprimatur of his own celebrity: Carmi the hero of Wingate's Special Night Squads, Carmi the daring arms thief, Carmi the defiant instigator of the flag mutiny.

But this evening his address took on an additional significance. After a busy day spent breaking camp and loading the big Dodge trucks, after a day spent preparing to leave Italy, the men understood they had come to something definitive. They were about to start a new era. A Jewish army, the gold Star of David on its soldiers' shoulder flashes as a symbol of pride, not submission, would soon be entering Germany as an occupying force. The extraordinary circumstances left the men racing with anticipation.

And now on this momentous evening, Carmi began to speak. He

wanted to share, he announced in a clear, unnaturally precise voice, the commandments for a Hebrew soldier on German soil.

He was not a public man, speeches did not come easily to him, so he read from a sheet that had been prepared by others; Maj. Shlomo Shamir, the Brigade's Haganah commander, was the principal author. But he was also not a trivial man. And despite his monotone, many of the men felt the emotion in his words.

"One. Hate the butchers of your people—unto all generations!

"Two. Remember: You are the emissaries of a people prepared for battle!

"Three. Remember: The Jewish Brigade in Nazi Germany is a Jewish army of occupation!

"Four. Remember: The fact that we come as a military unit with flag and emblem in sight of the German people, in its homeland—is vengeance!

"Five. Remember: Vengeance is a communal task. Any irresponsible act weakens the group.

"Six. Look like a Jew who is proud of his people and his flag!

"Seven. Pay them no mind—and do not come under their roofs!

"Eight. Let them be anathema—they, their wives, their children, their property and all that belongs to them—anathema for generations!

"Nine. Remember your mission.

"Ten. Your duty: Dedication, loyalty, and love for the remnants of the sword and the camps."

When he had finished reading, his voice nearly hoarse from the exertion, Carmi folded the paper and put it back into his shirt pocket. He was about to return to the ranks, when he called out to the troops in an angry, heartfelt voice: "Cursed be he who fails to remember what they have done to us!"

Thousands of voices, a unified and unsentimental chorus, responded, "Amen."

Moments later the Brigade was dismissed. They had been forced aside while the last battles of the war had been fought, but soon they would get a measure of revenge. Tomorrow they would head north, conquerors marching into the remains of the Thousand Year Reich.

The Star of David, by long agreement with the British War Office, was painted on the fender of the Brigade's olive green trucks in bright yellow. The next morning by the time the men had taken down their tents, packed their kit bags, and crammed into the backs of the trucks, other unauthorized markings had appeared on many of the vehicles. Written in tall chalk letters in German on the side panels and canvas roofs of the trucks were: *"Die Juden kommen!"* and *"Kein Reich, kein Volk, kein Führer!"* Large blue-and-white striped national flags hung from the back of the drivers' cabs. The effect on the soldiers as they passed truck after truck in the long line was cumulative and dramatic. A few disapproving officers, including Pinchuk, could not help wondering if they were provocations that would further stir the men's unsettled mood.

Nevertheless, no one ordered them removed. The slogans and the flags remained in place. And soon there was a cacophony of rumbles and growls as dozens of motors turned over, the trucks jerked forward, and the convoy moved out.

At first there was a festive air to the journey, and the platoons of men sitting side by side in the back of the trucks broke out in the martial "Battalion Song." As the line of trucks crossed the long pontoon bridge that straddled the Po, there was a proud sense that their new mission had begun.

But by noon the day had turned oppressively hot, and the trip became tedious and uncomfortable. The flat, green Italian countryside was lit by a scorching sun. The trucks unexplainably had slowed to a snail's pace—twenty kilometers, Peltz guessed with annoyance—as they crept along.

The men had removed their heavy outer coats, but the heat was inescapable. Crammed knee to knee in the back of the trucks, they began to realize the tightness of their quarters. The convoy moved with a haphazard, stuttering rhythm, and even sleep was difficult. A day that had started with clarity and in excitement faded into something that was merely annoying. Impatience and bad temper were the norm.

Yet by sunset when they made camp and drove their tent stakes into a spongy field, the men were more philosophical. Just being out of the cramped trucks seemed to lift their spirits. In four days, they reminded themselves, they would be in Germany. They could endure until then. *Die Juden kommen!*

That night, asleep in his tent, Peltz had a dream. He was mounted on Kary, the black stallion he had ridden as a boy, galloping across the fields of Zabiec in tandem with his mother and his big grandfather.

But in the morning when he woke, he knew it was an illusion; and much more painful, one that was impossible to hold. Kary, of course, was gone. An Arab bullet had prematurely ended his riding days. And Zabiec, his mother, his grandfather—he could not find in himself a solid reason for any hope at all.

Carrying this burden, he climbed back into the truck with his men. And the long journey resumed. *Die Juden kommen!*

After lunch the convoy started to climb up a steep winding road, and into the Alps. The stale summer haze that hovered above the dusty Italian plains was replaced by a clear, radiant blue sky. The air was crisp, its chill invigorating. And the vistas were magnificent: towering snow-capped peaks surrounded a verdant valley blazing with the bright, happy colors of wildflowers. Peltz felt there was something almost staged about the scenery, as if it had been self-consciously arranged for a photograph.

But as he continued to admire the perfect view, something began to intrude. Coming slowly down the mountain, heading in the opposite direction, traveling south on the same steep road, was another convoy. These trucks were smaller, painted brown. And seated inside were soldiers in field gray uniforms.

Peltz realized what he was seeing, and confirming shouts, loud and surprised, began to erupt from the Brigade's trucks: "Germans. They're German troops."

It was a convoy of Wehrmacht prisoners of war. There were dozens of trucks filled with soldiers and Mercedes sedans carrying officers.

The Brigade stared across the road. No one spoke. Every new moment pushed against the wall that held back their reserves of anger. The only sound as the two long convoys passed along the narrow mountain road was that of the trucks struggling on in low gear.

Then a soldier shouted in Hebrew: "They're laughing!"

It was the flaming match, and the fire it ignited raged instantaneously. The men went wild, aiming a volley of combat rations and shiny tins of jelly and margarine at the cowering German troops. They ripped up floorboards with their bayonets, and hurled the nail-studded planks. In some of the trucks the men found tools—stakes, picks, wrenches—and threw these heavy pieces of iron as if they were spears.

The shout of *Aleihum! Aleihum!* (Get them! Get them!) rose like a battle cry from the line of trucks. The furious barrage, an arsenal of flying metal, continued. Several of the Germans were bleeding profusely.

The British officers escorting the prisoners shook their fists at the undisciplined Jewish soldiers. One officer reminded the men about the Geneva Convention. These were unarmed prisoners, he tried to explain. But he was drowned out by jeers.

The Brigade trucks stopped short and the Jewish officers were formed into a line in the middle of the road. They were a human barrier between the troops and the convoy of prisoners. It was the only way the staff officers could hope to get the men to stop.

Pinchuk, shaken, shouted at his men, "What are you doing? They're unarmed. We're Jewish soldiers. We can't behave like this."

Peltz stood in the road with his arms at his sides, tall and imposing, as if daring any soldier to throw something. But he could not bring himself to say a word. He refused to criticize the men for something he wished he had initiated.

Carmi, meanwhile, sat stoically in the back of a truck with the men from his company. He was only a sergeant; it was not his job to stand in the road and protect the Germans. But he could not help feeling it was a demeaning outburst. More important, it was ineffective. You don't kill Germans with tins of margarine or even wrenches. And if you're not going to kill them, he saw no reason to bother.

Twenty minutes later the convoy of prisoners had passed, and the Brigade continued on. As daylight faded, they headed down into a wide valley. The trucks passed over an arched stone bridge and into the village of Tarvisio.

There was a mill house, its water wheel turning in a lazy circle through ice blue water. Beyond it was the village square, stores and houses made of plaster and white lime with dark wood roofs. In the doorways blond-haired children stood, curious and friendly, waving at the soldiers.

It was a perfect little Alpine town and Pinchuk, like many of the men, was charmed. He was glad they would be stopping here to spend the night. Soothed by this idyllic scene, it was easy to make light of the outburst that only hours earlier had filled him with such disappointment. It was possible to believe that it was of no significance, that it portended nothing, and that the commandments Carmi had delivered had not been irreparably shattered.

From the cab of his truck, he saw as they drove on a poster hanging in a store window. The words were somewhat faded, but the cartoonish drawing was clear enough. There was Uncle Sam, a hefty sack of dollars clutched in his hands, and a dog collar around his neck. Leading this beleaguered figure around on a leash was a hunchbacked unshaven Jew. He had a grotesque hooked nose and he was wearing a rich man's coat with an astrakhan collar: the picture of sinister, manipulative evil.

Pinchuk looked at this drawing, a poster in a store in a quaint mountain town, and anger coursed through him. If he had a rock, he would have hurled it through the glass window. The moment passed in a flash. Yet he felt a twinge of shame, of hypocrisy, as he recalled the indignant moralisms he had used to lecture the men. And he realized that what had erupted on the road was only a prelude.

But he could not allow his energies, his emotions, to be diverted when the Brigade entered Germany. No matter what happened, he had his own mission.

TWENTY-SIX

But the British would not let them go into Germany. The incident on the Alpine road was only a brief outburst, yet it was sufficient to make Command apprehensive about what might happen next.

The Brigade was ordered to remain at Tarvisio, to stand guard over a remote, mountainous corner of Europe where the borders of Italy, Yugoslavia, and Austria converged. Pinchuk, bitter and exasperated, complained to Peltz that their punishment was similar to the one God had given Moses for his rageful acts: They could see into the Promised Land, but they were forbidden to enter.

PRIORITY TOP SECRET
FROM: TROOPERS
TO: 21 ARMY GROUP, INFORMATION SHAEF FORWARD
REF NO THIS MESSAGE: 99135 SD2, 05 June 1945.

1. As result of public statements by Prime Minister in September 44 we are committed to use the Jewish Infantry Brigade Group at present in ITALY for occupational duties in EUROPE.

2. AFHQ proposed to use them in this role in AUSTRIA but in view of the many and complex

```
problems that have arisen in connection with this
occupation they do NOT consider it advisable to
add to them by employing Jewish Brigade Group
there in initial stages.
```

For much of the war, Tarvisio had served as an army convalescent depot, or *Lazarettstadt*, and the Wehrmacht's presence in the pretty little town remained strong. In fact, the beds of the large well-equipped hospital, a sprawling compound of buildings, were still filled with recuperating German soldiers. During the day a ragged ghost army of men in frayed field gray uniforms, an armless sleeve pinned to a shoulder, a wooden crutch to replace a missing leg, would hobble through the streets. The Brigade's headquarters was established in an abandoned SS army barracks. The men were billeted in houses that, weeks earlier, had been home to German troops; framed photographs of Hitler were removed from the walls. It was as if the raw smell of the Reich was still lingering in the air the troops had to breathe. The men found it oppressive and unsettling.

At the same time, the Brigade's official responsibilities in Tarvisio were minimal. They were assigned to guard the hospital (although from whom was never determined) and to march impressively in front of a massive munitions dump that had been assembled for a last stand that had never coalesced. The long June days were endless.

This was precisely what the War Office wanted. The Brigade had been pulled out of the final battles. And now that the war was over, the volatile Jews were hidden away in an obscure mountain village, contained and controlled until it was time for their demobilization back to Palestine.

But Tarvisio was not as unremarkable as the British had hoped—or as the Jewish soldiers had first complained. In the confused, unstable months after the war, Tarvisio had an unanticipated destiny. Its geography made it unique.

The village was positioned at the convergence of three national borders. Even more significant, it was only miles from the Austrian bridge that separated the British from the Russian forces. As a result, Tarvisio, an inconsequential clump of houses on the slope of a mountain, became a passageway across postwar Europe. It was a way station on the route west for thousands of desperate refugees.

Inadvertently, the British had positioned Jewish soldiers in a town whose two-lane highway, railroad, and mountain trails were the links between the ruins of the Reich and the promise of the future. It was not long before people, and news, began to trickle in.

First to arrive was a band of Yugoslav partisans, Tito's war-hardened guerrillas, and their welcome turned into a day-long celebration. Bottles of white wine that had been confiscated from the Wehrmacht officers' mess were raised in countless toasts. Grinning horsemen galloped through the streets, leaping recklessly over high fences and daring the Palestinians to equal these feats. Peltz longed for the days when he might have given them something to cheer. Instead he consoled himself by dancing a polka with a hearty red-cheeked woman, trying not to be too concerned by the grenades bouncing from the bandolier cutting across her chest.

Soon, though, other visitors arrived with more disturbing information. These first intimations of the magnitude of what the Nazis had done were so incredible that the men found them difficult to accept. Peltz knew there had been pogroms. He had read in the newspapers about the labor camps. Still, the stories he now heard seemed impossible. He did not want to believe them.

So when Peltz and a group of men from the Brigade walking near the railway station discovered some Ukrainian refugees crouched around an improvised stove, an inverted helmet filled with a thin stew bubbling over a fire, the conversation that took place seemed more ludicrous than ominous. "We're from the Jewish Army," a Czech-speaking soldier had explained. The refugees laughed. That was, they merrily agreed, quite a good joke. "I saw the Jewish Army," one refugee teased back, her hand waving to the sky. "They went up in smoke."

Or there was the time when Pinchuk and some friends went into a café and, despite the nonfraternization law, were quickly joined by a trio of young Austrian girls. "Where are you boys from?" one of the girls asked. "Palestine," Pinchuk announced. "We're soldiers in the Jewish Army." The girls laughed. "Don't pull our legs. There aren't any more Jews except in horror stories. They don't exist anymore. They're all gone."

But these exchanges could no longer be dismissed as the ignorant comments of peasants and schoolgirls when the first ragged Jewish refugees finally found their way to the Brigade. The soldiers now saw the proof with their own eyes: European Jewry, centuries of civilization, had been systematically reduced to these sad remnants.

First there were four young Jews from Poland, boys in their late teens dressed in stolen German army uniforms, who had fled from a camp at Ebensee, near Salzburg, in the American zone. Then there was a group of Hungarian tailors who had been kept alive in the camps only because they were craftsmen; toward the end of the war they had been handed over to the SS, but, another miracle, they had been liberated in upper Austria by the advancing Red Army. And there was a blond woman, well dressed, attractive. In a flat, emotionless voice, she admitted to a large circle of soldiers that she had survived the war in Austria by pretending she was a Christian. In the same unnatural monotone, she explained what had happened to those who were not as inventive. She told the men about the roundups and the death camps, how the dark smoke from their brothers' burning bones polluted the sky day after day. By the time she had finished, the soldiers were in tears. Even Carmi, usually steady and impassive, was sobbing openly.

Each day there were new arrivals, and each day brought more reasons for tears. The disheartened soldiers tried to imagine what had happened to their relatives—mothers, fathers, sisters, brothers—but when they did, their imagination would stall. There was no hope.

Confronted with this evidence of the systematic destruction of the Jewish people, Pinchuk simply gave up. He admitted his failure, and acknowledged his cowardice. There was no point in looking for his family. What sense was there in sifting through the charred remains? Defeated, he retreated deep into his private world, surrendering himself to the pain of his memories; and resigning himself to the fact that his own salvation was now impossible.

Peltz, too, understood there was little logic or reason for hope. But he felt he had to go to Poland. He wanted to walk through the fields of Zabiec. He needed to convince himself that he had done whatever he could, that he had tried. Perhaps then he could relinquish the past and move on.

Carmi, however, was not tempted to look back. With a soldier's stoic acceptance of death, he lived with his family's certain destruction. He was saddened, yet he was not defeated. He did not focus on the lost ghettos of Europe. He imagined the shining cities he would help build in Palestine.

But he also was a pragmatist, and he knew that it was vital for the Yishuv to know the dimensions of the horror in Europe. The information had a political value: The nations of the postwar world would have to acknowledge what had been done to the Jews; and they would have to make amends.

When Peltz announced that he was going to cross into Austria and then on to Poland, Carmi agreed to come along. He did not concern himself with the army regulations that would be broken by their unauthorized trip. He did not worry that the borders were patrolled by military police, or that soldiers were forbidden to travel without proper papers. He simply said that he would find a jeep. He wanted to see what was out there with his own eyes, and report back to the Yishuv.

TWENTY-SEVEN

The war had ended only a month ago, and the Allied MPs were not very rigorous. Peltz's officer's uniform and confident demeanor were sufficient authorization at every checkpoint. With Carmi at the wheel, they traveled through both the British and American zones without being questioned. And as they drove on, they tried to prepare themselves for what they would find.

They arrived at the gates of Mauthausen concentration camp on a bright June morning three days after leaving Italy. An American sergeant led the two men around, and it did not take Peltz long to realize there was no possible way he could have been prepared.

He had never thought much about concepts like good or evil. As a soldier and as a man he acted intuitively, yet decisively. It had never been necessary to define the universe as a battleground between the forces allied with a white-bearded God and those of a pitchfork-wielding Devil. But walking through Mauthausen, Peltz found he needed to categorize what he was seeing. He needed to make a connection to a concept he could fit into his imagination. Otherwise it was all too difficult; and he feared he might lose control.

He told Carmi what they were witnessing was "the Devil's work."

The Devil was in the washing room—as clean as a hospital

surgery with its cluster of steely showerheads that spewed Zyklon gas, not water.

The Devil was in the row of crematoriums—nearly thirty buildings, each with its own furnace and huge, powerful fans that would stoke up the flames until they burned like the fires of Hell.

The Devil was in the vindictiveness—the brick foundations had been laid for fifteen, or maybe twenty, Peltz was in no mood to count, additional crematoriums. A sustained inferno of hate.

The Devil was in the haphazard stacks of gray skeletal corpses cooling in an improvised morgue—they had somehow survived the war, yet they could not escape. Liberated, they had succumbed to typhoid, malnutrition, to disease.

Wherever he turned, Peltz heard the Devil's voice hissing in the air—Welcome to your fate, Jew.

Later, walking back to the Mauthausen gate, Carmi and Peltz noticed a crowd of survivors—perhaps a dozen reed-thin men, some in the hospital pajamas the Americans had issued, others still in black-and-white striped prisoner outfits—surrounding their jeep. From the distance, it appeared that something about the vehicle had caught their attention. Then, as the two soldiers got closer, they saw that the crowd was examining the dust-covered Star of David on the bumper of the jeep.

Carmi and Peltz went up to the survivors, and the group grew quiet. They stared apprehensively at the two men.

"Don't be afraid," Carmi said in Yiddish.

The survivors still did not speak. Peltz felt guilty—of his health, his strength, his good fortune to have been spared.

Carmi tried again. "We're Jews," he said.

Confused, a man pointed to the Star of David on Carmi's sleeve. He asked hesitantly, "You're Jewish angels?"

No, Carmi explained. They were just soldiers. Soldiers from Palestine.

The survivors could not believe it. It was too fantastic—a Jewish army. They needed to make a link to something that they could comprehend. Like Peltz, who had believed that only the Devil himself, not mere men, had been capable of creating such evil, the survivors found it easier to imagine that angels, not Jewish soldiers, were standing in front of them.

Peltz did not speak Yiddish and was unable to follow precisely what had been said. But he instinctively extended his arm as if to shake hands.

A frail man stepped toward him and touched the tip of his finger tentatively to the skin of Peltz's outstretched hand.

It took Peltz a moment to realize what the man was doing. He needed to discover for himself if Peltz really was made of human flesh, not something supernatural.

Tears fell from Peltz's eyes. He cried for these survivors, for his parents, for his grandfather. He cried for the Jewish people. He reached out and hugged the thin, bony man to his chest. Peltz held him, and the two men mourned together, bounded by their deep shared sorrow.

Carmi returned to Tarvisio. He had planned to go to Poland with Peltz. He had thought that he would try to learn what had happened to his parents and his sister, Ella, and her children. But after what he had seen and experienced in Mauthausen, he realized that he knew enough. And spurred by all he had seen, an idea was already beginning to take shape. He would use his time for other matters.

Peltz, however, had to go on. He had long ago discarded his vision of a triumphant return as a British officer. There would be no celebrations. No one would be waiting at Zabiec. But the force of a genuine compulsion was driving him. He had to see.

Traveling with a vaguely worded letter he had obtained from a friendly official at the Russian delegation in Linz, he arrived by train in Kielce, where his father had been the director of the Jewish Hospital. As a result of this position, Johanan learned, Dr. Peltz was among the first Jews shipped to Auschwitz. Peltz's mother was sent to Treblinka. His grandfather had escaped into the forest surrounding the small town of Bochnia. He hid with his new wife and a group of Jews in a bunker they had constructed out of trees and then covered with leaves and moss. He survived in this way until the Germans, acting on information supplied by a Polish farmer, came hunting for them. According to the story Peltz heard and very much wanted to believe, his grandfather had shot two soldiers before being cut down by a burst of automatic gunfire.

As for Zabiec, it, too, only existed as a memory. The Communists had already claimed the land and begun distributing it to the local peasants. The big house, with its well-furnished rooms attended by a company of servants, gardeners, and cooks, was a ruin. When the Russians had crossed the Vistula, the mansion had been caught in the artillery fire between the two armies. It was totally destroyed.

Peltz walked through the rubble of bricks and stone. The only thing he found intact was a brass door handle screwed to a cleaved and burned door. He used his pocket knife and, after considerable effort, he pried the handle from the door. He slipped the tarnished piece of brass, his inheritance, into his pocket and suddenly he was in a hurry to leave. Zabiec no longer held any interest for him. He wanted to return to the Brigade, to his people.

TWENTY-EIGHT

Leah waited for her punishment, but an entire week passed and there were no recriminations. When the Russian officer finally came, he put her in a horse cart and drove to a train station. The war was over.

Leah returned to Reflovka. Looking at the village for the first time since she had fled three years earlier, she had no illusion that this was still her home. While outwardly familiar, the streets and houses were now as alien to her as those in the towns she had glimpsed through the train window.

But she needed somewhere to stay, and she did not have any money. She reluctantly went to the house she had shared with her parents and brother.

A Russian soldier standing by the door demanded, "What do you want?"

Leah suspected that he had been drinking, but she tried to explain her problem to him. "This is my house," she said. "I lived here before the war."

"Perhaps," he said. "But it is ours now. Go away."

"Where can I go?"

The Russian considered a moment, then said, "Go away before you're sorry."

* * *

Leah walked down the main road, peering into windows. She needed to find a room where she could stay, she needed a way to make money to buy food. A light from a basement window spilled out into the street. Leah, looking in, saw a Jewish woman hunched over washing clothes. Leah did not recognize her, but she knocked on the door.

"Have any Jews come back?" she asked. "Have any survived?"

There was a family living down the street, the woman said. A father and his two daughters had escaped to the forest, hidden in another village during the war, and now had returned. They were living in the house next to the synagogue.

"That's Simcha Bert's house," Leah said.

The woman shrugged. She was a stranger to Reflovka.

Leah thanked her and hurried away. The possibility that her friend and classmate Pesel had survived filled her with excitement. Before the war Pesel had visited so frequently that she had been considered part of the Pinchuk household. Pesel would tell Leah with gratitude, "Your mother is my mother."

Leah knocked on the door. A man with a long scraggly white beard answered. The last time Leah had seen Simcha, he had been dark-bearded and big-bellied, a generous smile animating his round face. Now he resembled a man in the last years of a long, difficult life.

"Leah?"

Before she could answer, he was hugging her. And he asked through his tears, "Your parents?"

Leah could only sob and shake her head.

"Then you will become my daughter," Simcha said. "You will stay in my home."

Simcha had been a butcher before the war and he quickly returned to his old work. He brought home meat, vegetables, fresh bread. The familiar smells that filled the small kitchen reaffirmed that reason had been restored to the world. Leah slept in a bed with a mattress.

Simcha's two daughters were her two sisters. She had people to talk to. "Your hair's growing out," Pesel told her. "It's such a pretty color. You'll look beautiful." It was almost possible to believe things had always been this way.

But life in Simcha's house, while orderly, reasonable, and even sweet, was only an interim existence. She had made a promise to her father, and it was the plan for her future.

At the end of her second month, Leah announced that she was moving to Rovno. It was a big city. She would get a job. And somehow find a way to get to Palestine, and her brother.

TWENTY-NINE

Two women had been attacked. In the middle of the night, three soldiers broke into a house in Tarvisio and assaulted a mother and her teenage daughter. The women would have been raped, the soldiers were all over them, but their screams scared them off. The men grabbed a handful of bills, took two watches from a drawer, and fled.

The women said the German-speaking soldiers had boasted they were Jews settling scores. Immediately, the British MPs suspected the men were from the Brigade.

An identification parade was ordered and the soldiers billeted near the house were lined up. The women stopped in front of every man, looking him up and down, trying to identify their attackers. The outraged mother wore a black skirt, black stockings, and a kerchief over her head. Her shy, blond daughter was quite pretty. When the women were done, they conceded to the MPs that they could not find the faces they remembered.

The investigation was terminated. And neither the women nor the MPs ever discovered that three soldiers from B Company had been missing from the parade, their names and places in the ranks assumed by men from another battalion.

This was the disturbing news that greeted Carmi on the morning he crossed the Austrian border into Italy, and returned to camp.

Later that day Carmi took a long solitary walk. He followed the stream that bisected the village and then continued up along a mountain path. He needed to sort things out in his mind.

He understood what had happened. He understood the urge, the temptation, to strike back indiscriminately at the people who had so complacently allowed the Nazis to do as they wished. The reasons for the shameful attack, same as for the incident on the mountain road with the German prisoner convoy, were clear enough.

He could even imagine how it might have transpired. Three Jewish soldiers sitting around at night, talking, drinking. A sudden, offhanded suggestion. And then a deep, residual anger would allow a man to surrender, to become passive, to just go along—until he was caught up in something he had never anticipated.

After the horror he had seen in Mauthausen, how could he criticize any impulse for retribution? Perhaps this was an inevitable—even justifiable—response to a community that had fostered such evil.

Now was the moment for the Jews to turn the tables. To repay not just the Nazis, but all the centuries of persecution. Perhaps the time had finally come for Jewish soldiers to unleash all restraints. For Jewish Cossacks to chase through the ruins of Europe, raping, pillaging, and avenging. In his present mood, it was an appealing image.

And this terrified him.

Because at the same time, Carmi knew that wild, indiscriminate vengeance was not only dishonorable, it was wrong. He could not allow himself or his men, Jewish soldiers, to be drawn in. To find the sanction for such unbridled behavior—attacking women! stealing watches!—in the death camps was an insult to the memory of those who had been murdered.

Yet something had to be done. He was convinced it was the passivity of the ghetto Jews, their reasonableness, that had led them to the crematoriums. Jews needed to understand that there was power in their collective rage. And the anti-Semites had to learn—and fear—that Jews would strike back.

But Carmi also knew any action needed to have meaning for it to be valid. It had to possess a moral force. It had to be disciplined. It had to be a weapon employed in a soldier's way.

His mind was racing, but his thoughts had begun to clarify. By the time he returned to camp, he was closer to deciding what he would do. First, however, it was necessary to speak with Shlomo Shamir. He would meet with him that evening.

"Sit, Israelik."

This greeting was a signal. Maj. Shlomo Shamir's words told Carmi that it was safe to discuss Haganah business. An order of "At ease, Sergeant," would have been a warning, and Carmi would have spoken only of Brigade matters.

But now that their covert discussion had been sanctioned, Carmi made his request. "I want to be assigned to the Brigade Intelligence Unit," he declared without preamble. He let his words hang in the air. He did not feel an explanation was necessary.

The two men respected each other, but they were not friends. Shamir considered Carmi too adventurous, even intemperate. Carmi felt Shamir was too cautious, even timid. Each thought the other personified all that was wrong with the Haganah.

And so there was no further discussion. Maj. Shamir simply said that he would prepare the necessary paperwork. Then Carmi left.

The transfer was quickly processed by Command, and Carmi threw himself into his new assignment. With the help of Robert Grossman, an Austrian-born soldier in the First Battalion who was already working in Intelligence, he sifted through all the diverse information that had been sent to the unit. Fortified by cups of strong British tea, the two men spent days and night reading stacks of single-spaced typed pages—rumors, field reports, classified Allied intelligence summaries, top secret communiqués.

The unsorted information was often compelling ("It was Gen GREINER, an ardent Nazi and an aggressive Comd, who is reported to have said that the Div would continue to fight until it could be fed from a single Coy mess . . .") and the rumors were often preposterous (". . . a woman alleged to resemble Eva BRAUN, mistress of A. HITLER, was seen at the KLAGENFURT railway station . . ."), but the two men could not be diverted. They searched the pages looking for one common thread—the names and locations of former members of the SS and Gestapo.

The Americans were eager to help. Trading on both his official assignment with a British intelligence unit and his membership in the Jewish Brigade, Carmi marched into the headquarters of the American 301st Intelligence unit and recruited an army of volunteers. More than half the unit were German-speaking Jews, many had relatives in the death camps, and they opened their files and shared their carefully researched lists.

They also offered Carmi practical hints: To confirm that a suspect served in the SS, look for the blood group tattoo beneath the armpit. Refugees who appeared too healthy, suspiciously fat and bright-eyed, might very well have spent the war in a concentration camp—but as guards, not prisoners. A lieutenant from the Bronx treated Carmi to an elaborate dinner and told him, "It's like fishing. You find yourself a minnow, and you use him as the bait to catch yourself a whale."

But this was only the paper chase. Once Carmi felt he was beginning to have a grasp of the information—who was at large, where they might be hiding, what networks had been established in the last days of the war to facilitate the escapes—he was ready to go into the field.

He did not have to go far. Tarvisio, despite its modest size, was a community of large secrets. For much of the war the SS had billeted a full battalion in the town. Gestapo officers supervising the deportation of Italian Jews had worked out of the local railway station and the one in nearby Klagenfurt, Austria. Many of its citizens had been employed by the German officials who had controlled the region.

Refugees from all over Europe were pouring into the area. And the hospital complex was filled with convalescing Wehrmacht soldiers and German doctors. Carmi had no doubt that many of the answers he was looking for were hidden in the town. It was simply a matter of getting people to talk.

Carmi was a natural interviewer. He was patient, polite—and when necessary, completely intimidating. He also knew about lies. In his long secret life in the Haganah, Carmi had invented hundreds of stories, fabricated his way through countless unnerving situations. He was attuned to all the tricks. He could pick up on a hesitancy, a smile that was too ingratiating, an explanation that seemed too careful and calculated. He was rarely fooled.

He would deferentially offer his British identification card, ask questions, and listen with only the occasional interruption. The hospital quickly became his favorite site to mine. It was the hideout, he was convinced, of dozens of SS men feigning illness. He was always concerned, tactful, and respectful of the condition of the patients. He let them go on vaguely, sometimes disingenuously, and when they pleaded that they were getting tired, he apologized for keeping them so long. But Carmi filed everything away.

And at night he returned. He would revisit the home of the apologetic man who had said he could not remember the name of his German supervisor at the train station. He would return to the hospital bed—night after night the guards would pretend not to see him—where the recuperating soldier had a bandage stretched across a "wound" near his armpit. And he would begin his interrogations.

At the hospital, Carmi would wake his suspects at three in the morning with a slap to the face and a torch shined in their eyes. Standing over the bed, waving an intimidating fist, he would start his questioning at precisely the point they had begun dissembling. All his former patience, his careful politeness, belonged to another role. Now he would yell in his booming sergeant's voice and demand answers. If anyone doubted that he was a serious man, Carmi explained how dangerous it would be not to cooperate. The threat of a revolver butt smashed against a broken bone was usually harrowing enough. Only rarely did he need to prove the pain was excruciating.

In this way, he got answers and information—names, locations, aliases of men and women who had actively conspired to exterminate the Jews of Europe.

Yet even as he was making his list, he still had not fully determined what he was going to do with it. Part of him wanted only to pass on these names to the authorities, to make sure the war criminals he had located were brought to trial. Men would be punished, and justice would be done.

But all the time that he hunted, every moment that he spent gathering names and locations, he was also aware of another alternative, another way to get revenge. And one incident forced Carmi to choose.

As he was collecting intelligence, two sources told the same intriguing story. First Carmi heard it from a German surgeon at the hospital. The doctor had allowed several SS men to hide in his ward, and now he wanted to make a trade.

Forget about me, he begged during one of Carmi's late-night interrogations. I can give you something much bigger. There's a couple living near the border, in Austria. The man was a Gestapo official, a supervisor of the racial deportations, very high up. His wife, though, she's the one you really want. She was in charge of the confiscation of Jewish property for all of northern Italy.

Carmi, suspicious, demanded to know how the surgeon knew this.

I saw it with my own eyes, the surgeon insisted. I went to their house for dinner and they let me look. There were chests full of jewelry and gold. They gloated. They were proud of the booty they had "requisitioned" for the Reich.

In a café in town Carmi soon heard a similar story. The man who told it was angry and did not have to be threatened. He could hardly wait to share his information with a British intelligence officer.

But it was not the injustices that had been done to the Jews that troubled this man. It was the inequities of the post-war world that left him full of rage. The war had reduced everyone to poverty, he complained. Meanwhile, a selfish couple across the border in Austria were hoarding a fortune. "Somebody," the man said, "ought to do something about it."

THIRTY

It was not quite midnight when Carmi and Grossman arrived at the well-kept three-story home just over the border in Austria. Carmi knocked thunderously and repeatedly on the heavy wooden door. It was meant to be rude and intimidating. At the same time he called out in German, "British Intelligence. Open this door immediately."

He kept pounding on the door until a light came on in an upstairs window. Moments later, the door was opened.

"What's this all about?" The man had a clean shiny face, very pink, and he wore a robe over a pair of blue pajamas. His thin gray hair was askew. "Do you know what time it is?"

Carmi shoved through the open door, and slammed it behind him. "There are some questions I want you to answer," he said.

The pair of armed British soldiers stood deliberately close to the middle-aged man, but he was calm, even defiant. He looked at the intruders contemptuously.

"By what authority are you here in the middle of the night?" the man challenged. He told them to come back tomorrow morning and he would meet with them at a decent hour.

Carmi pointed to the gold Star of David on his shoulder. "Authority? *Ich bin Jude.*"

The man's head dropped in an unconscious gesture of surrender.

When Carmi ordered him to get his wife, it was an old man who obediently obeyed.

Once the interrogation began, however, the man managed to steady himself. If all the two Jews were going to do was ask questions, he was not worried. He was ready.

And, Carmi began to realize, it had been a mistake to seat the wife next to her husband. He had hoped that by keeping them both on the sofa, he could watch the pair while Grossman searched the house. But he had underestimated the wife.

She was a thin, sour-faced woman with a neat bun of silver hair. Her contempt was undisguised. She refused to look Carmi in the face and she made it clear that she would not cooperate. Her defiance emboldened her husband.

Whom did you work with in the Gestapo? Carmi barked. Where are they hiding? Where is the jewelry, the gold, you stole from the Jews? I know you have records, show me!

But the man's previous fear had vanished. *"Ich weiss nichts."* I know nothing, he repeated flatly.

Grossman's search also proved futile. Each time he returned to the living room, he had the same report: "Nothing."

Finally, Carmi realized there was no point in asking any more questions. It was a stalemate, and Carmi knew it was up to him to determine how it would end.

In the long silence, the ticking of the ornate clock on the mantelpiece was the loudest sound in the room. As if he were alone, Carmi's mind began to wander. *Jewish angels?* the surviors had asked. *No,* he had answered, *Jewish soldiers.* And goaded by this memory, Carmi finally acknowledged what he had been setting out to do from the moment he had decided to make his list.

"If I find anything in this house to connect you to the Gestapo, to crimes against the Jews," Carmi said, "I will kill you."

The search continued for eight hours. They inspected every room, then every drawer in every room. A loose floorboard was pried up. Grossman's rifle butt destroyed a closet wall. They found nothing. But then Grossman called to Carmi from the kitchen.

Carmi led the couple at gunpoint into the yellow room. In a corner was a fieldstone fireplace large enough for a man to walk into. The fireplace was laid with logs. A carpet of black ashes was spread beneath the wood. And Grossman was reaching between the logs, poking into the ashes.

Like a magician, he pulled a khaki-colored canvas bag out of the fireplace. Then another. And another.

"I never saw those before," the man insisted.

His wife turned to leave, but Carmi grabbed her arm. "Don't you want to see what we found?"

Inside the first bag was a collection of jewelry—heavy gold necklaces, diamond rings, brooches encrusted with colored stones.

The second bag held bound stacks of paper currency.

The third contained several Luger pistols.

Carmi's reaction was one of relief. All night, he had feared that his rage would push him to kill the couple before he had proof of their crimes.

The couple started to cry. Carmi, unaffected, reached for his pistol. In his mind they were already dead.

"Don't kill me, don't kill me," the man begged.

"Please . . ." the woman cried.

Carmi told Grossman to line them up against the wall.

"Wait," the man said. "I can help you. I'm nothing. But I know names. I can help you. I can tell you about important people. Colonels. Generals."

"You're lying," Carmi said.

"I'm not. Please. I can give you a list. A long list. I have records."

Carmi fingered the revolver's trigger and stared at the couple pressed against the wall. He wanted to get this over with. Yet he also remembered what the American lieutenant from the Bronx had told him. Find yourself a minnow, then use him to catch a whale.

Carmi lowered the pistol and said, "Let's see your list."

"You won't kill me?" the man tried to bargain.

"Let's see your list."

The man sat at his desk in the room he used as an office and referred to several notebooks as he typed. He worked for five hours. When he was done there were eighteen meticulously typed pages, complete with names, dates of birth, addresses, physical descriptions, and personal histories. Each name identified an officer in the SS or the Gestapo. But this list, the man assured Carmi, was only the first installment. If the soldiers promised not to kill him, he would provide them with new pages each week filled with additional names.

It required a great effort, but Carmi agreed.

A day later Carmi was still pretending to himself that he had not decided what to do with the list. He tried to convince himself that he was seriously considering giving the names to the Allied intelligence unit. He tried to persuade himself that war crimes trials would be sufficient vengeance. But he also knew that just yesterday he had been prepared to execute two people.

Still, killing two people was different from setting out to track down and murder hundreds. The list of names went on for eighteen pages! A team of men would be necessary. On this scale, the logistics were daunting. It would be dangerous, even reckless. And in its vast cold-bloodedness, maybe even wrong.

In the end, though, Carmi decided that the risks were irrelevant. And the morality of what he was contemplating was unimpeachable. Jewish honor required Jewish vengeance.

He made a second list. Unlike the first, this version contained the names of only minor Wehrmacht functionaries. Carmi gave it to his superiors in the intelligence unit. The names he did not share were all senior Nazi officials and SS men. Carmi had other plans for them.

THIRTY-ONE

In Rovno, Leah found work. And she fell in love.

On her first day in the city, she met Katz. She was walking by a café and he called out to her. "Good day," he said and tipped his hat. Leah ignored him, but he left his table and followed her down the sidewalk.

"Please, let me buy you a coffee," he persisted. "You're Jewish, right? What could be the harm? A Jewish boy buying a cup of coffee for a Jewish girl."

Leah was exhausted from her train ride. And the coffee was free. When she turned and went back to the café, she allowed him to carry her valise.

"So why are you here?" Katz asked. "Family? A boyfriend?" He had already convinced her to have a slice of black bread and jam with her coffee.

Leah explained that she was looking for work.

Katz leaned across the table and whispered, "I'm in the black market."

Leah laughed. That was not the sort of work she was looking for.

But once again Katz was persistent. Where else was she going to get a job? Where else was she going to make such money? There was really little risk. He worked with a Russian lieutenant who had the use of a military pickup truck.

"I drive the truck to Lvov, the lieutenant sits next to me, and nobody bothers us. Join us," he offered grandly.

She stayed with a family her parents had known, a barber, his wife, and their two children. They lived in one room and slept on folding cots. In the morning, they closed the beds and the barber used the room as his shop. And Leah left to make her purchases. Using money Katz advanced her, she went around Rovno buying the merchandise they smuggled into Lvov.

Twice each week they made the four-hour trip, Katz driving the entire way without a stop. Leah sat beside him, her bag bulging with packs of cigarettes and tins of baking soda. The Russian lieutenant sat on Leah's other side, a smile on his broad handsome face, his elbow leaning carelessly on the window.

Leah and the Russian would talk for the entire ride. He was polite, and his voice was gentle. He was interested in whatever she said. His name was Aleks, he was twenty-four years old, and he had shimmering blue eyes. She began to look forward to these trips.

Leah was delighted to discover that she was quickly earning more money than Katz had promised. She even considered buying new clothes. The black dress Peshel had given her was becoming shiny from wear. But, she told herself, it was not ripped. The more money she could put away, the sooner she could afford a ticket to Palestine. What did it matter what she wore?

One afternoon, though, she went out and bought a pair of high black boots, a robin's-egg blue skirt, and a blouse with a wide rounded collar. At night she folded the skirt into thin pleats, and then slept with it under her mattress. In the morning when she pulled it out and put it on, she thought she looked very fashionable.

When Aleks told Leah he liked her new skirt, she went out and bought two more.

* * *

As they sat side by side in the bouncing truck, Aleks sang to her. He had a soft, warm voice, and when he fixed her with his bright blue eyes, the words penetrated her heart.

> The light blue colored scarf
> Has fallen from your long thin shoulders.
> You promised me you would not forget
> Our evening and our time together.

When they went to the market in Lvov later that day, their hands were linked together.

Aleks told her he lived in Moscow. He was an only child. When the army released him, he would reapply to the university. His father was a professor so he was confident there would not be a problem. Aleks wanted to be an engineer.

Leah felt she could trust him completely. The Communist government prohibited Jewish emigration, but she told him about her brother in Palestine and her plan to join him.

"How will you go?"

"When I have enough money, I'll travel to Italy, or perhaps France. I'll find a boat."

"You told me, that is all right. But don't tell anyone else," he said.

What he meant, Leah realized, was that she should not tell anyone who was not Jewish. What he meant, Leah understood, was that there was no future in their romance.

But when Aleks announced he had been ordered home, the thought of their being apart filled her with an unexpected sadness. The lieutenant was also surprised by the emotions swirling in his own heart. He asked her to marry him.

"It's impossible," she said.

He pleaded with her to come to Moscow. His father could get her into the university. They could make a life together.

"It's impossible," she repeated.

Think about it, he begged. He would be leaving in three days.

Leah considered his proposal. It was an opportunity that would finally deliver her from loneliness and danger. She wanted to put herself in his care, under his protection.

But what about the pact she had made with her father?

Where does my future belong? she asked herself.

Three days later, Leah again told him no.

As they said good-bye, she asked Aleks for one last kindness.

"When you get to Moscow," she said, "I would be grateful if you would send my brother a telegram. Tell him his sister Leah survived the war, and that she is coming to see him."

Palestinian Police Sgt.
Johanan Peltz, the hero of Sdom.

J. PELTZ

Israel Carmi, with Wingate's special night services, the troops charging into battle to the sound of the shofar.

I. CARMI

A. PINCHUK

Right: The Pinchuk family (Reitze, Meir, Arie, and Leah) in Reflovka, the town "as big as a yawn."

Below: Arie, the student who made the daring escape to Palestine.

Podpis

Pinczuk Lejb

Rok szkolny 1933/4

Legitymacja ważna do 20 sierpnia 1934 r.

Pinczuk Lejb

jest uczniem klasy siódmej

Pieczęć

Dyrektor

Rok szkolny 1934/35

Legitymacja ważna do 19 sierpnia 1935 r.

Pinczuk Lejb

jest uczniem klasy ósmej

Pieczęć

Dyrektor

A. PINCHUK

BETH HATEFUTSOTH

Above: Brigade recruiting in Tel Aviv.

J. PELTZ

Right: Charging up this hill near the south Dead Sea, Sgt. Peltz received the wound that ended his riding days.

J. PELTZ

The Hartuv police station. When the British officer complained about the Star of David above the door, Peltz (center) knocked him down.

A. PINCHUK

Leah Pinchuk, a twelve-year-old afraid to sleep without a candle burning next to her bed.

A. PINCHUK

Reflovka's Jews were led at gunpoint down
this road to the pits that would become their graves.

J. PELTZ

Second Company races forward.

J. PELTZ

J. PELTZ

Left: Taking position, waiting to attack.

Below: Lewis gun training, Second Company.

CENTRAL ZIONIST ARCHIVE

Above: Brigade marching to the front, March 1945. At last they would get to fight.

CENTRAL ZIONIST ARCHIVE

Right: The brigade's artillery fired into the German positions across the Senio River. The shell reads *Greetings to Hitler*.

J. PELTZ

Captain Peltz led the bayonet charge into La Girogetta.

The funeral of Haim Brot, killed as he defused mines in the attack on La Girogetta.

J. PELTZ

J. PELTZ

Brigade engineers preparing the way across the Po . . .

J. PELTZ

. . . and into the remains of the Thousand Year Reich.

BETH HATEFUTSOTH

Above: Brigade soldiers by the crematorium in Bergen-Belsen: "The Devil's Work."

A. PINCHUK

Right: Capt. Arie Pinchuk searches for his sister Leah.

CENTRAL ZIONIST ARCHIVE

Brigade convoys—Stars of David on the doors of their trucks—crisscrossed Europe to rescue the survivors.

CENTRAL ZIONIST ARCHIVE

The Brigade raced into Graz, brought out 1,000 survivors, and helped them get to Palestine.

Right: Sgt. Israel Carmi led the *huliyots* in vengeance before devoting himself to the rescue of his people.

I. CARMI

Below: Teaching the children, preparing them for their new lives.

BETH HATEFUTSOTH

BETH HATEFUTSOTH

"They treated us as if we were their brothers, their sisters, their children."
Party for the Jewish children found in monasteries.

BETH HATEFUTSOTH

Crowded onto the ship at La Spezia, Israel Carmi, "boat commander."

BETH HATEFUTSOTH

"Let My People Go"—the hunger strike at La Spezia.

A. PINCHUK

Leah, no longer alone, on the way to Palestine.

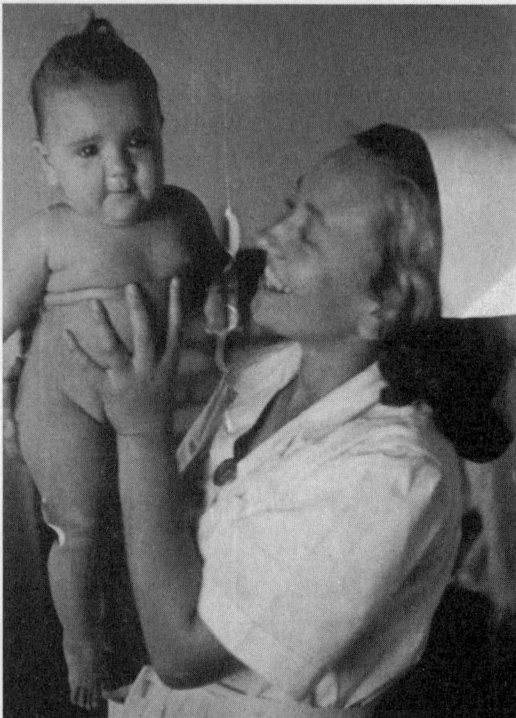

A. PINCHUK

Left: Leah's new life: a nurse in Israel.

Below: Arie, Leah, and Boris and his wife, reunited in 1985.

A. PINCHUK

THIRTY-TWO

As instructed Peltz parked his jeep by the road, but he had trouble finding the path leading to the farmhouse. He searched about in the dark, but it was no use. He wished he had brought a torch. At last he arbitrarily picked a spot and began walking up the rocky hillside.

The moonlight was thin and unhelpful. He knew he was only two kilometers away from the main camp in Tarvisio, but walking through the tall grass, Peltz could not help feeling he had entered into a completely foreign place.

Carmi had told him the owner of the farm in Camperosso had committed suicide before the war, and the house had been abandoned ever since. Peltz had agreed that it seemed like the perfect place for a secret meeting. But in the nighttime the isolation was disconcerting. And he was already on edge. Peltz heard a noise and he immediately tensed. If it was an MP, that would mean they had been betrayed. Many of the men believed there were spies in the camp, Jewish soldiers who routinely reported to the British. Peltz had always assumed this was true, but it had never been his concern. Let the Haganah boys worry about it, he had once told Pinchuk. But back then he never had anything he needed to hide. When he saw that it was only some small animal rustling through the underbrush, a rabbit perhaps or a night-bird, it was a great relief.

He reminded himself that there had been nights not long ago when he had to wonder about the Germans hiding in the shadows, not the British. On those missions he had also been uneasy. That first night in the woods on the way to La Giorgetta, he had gotten lost, too. But tonight there would be no medals, no glory. Only a court martial if he was caught.

Yet he never considered returning to his jeep and driving away. The memory of what he had seen at Zabiec overwhelmed every doubt, every misgiving, every uncertainty. The moment of resistance, when he might have been able to look at what he was setting out to do with a sense of objectivity, had long passed. In his mind he had murdered his parents' and his grandfather's killers so many times that it might just as well have been an actuality. It was Carmi's idea, but Peltz had been waiting, prepared and eager.

Cautious but determined, he worked his way up the steep hillside.

Peltz was late. He never found the trail and had to climb over a fence that circled an overgrown turnout. When he finally reached the farmhouse, the other men were already there.

No one spoke to Peltz when he came in. There were only a few curt nods. Even the air in the room felt thick, as if weighed down by the seriousness of their undertaking. There was no room for distractions.

A lantern hung from a beam and there was an open bottle of cognac on the wooden floor. Waiting in the tense silence, the room lit by a soft, diffused light, Peltz found he recognized many of the faces: Meir Zorea, "Zaro" everyone called him. He was the tough Russian who had won the Military Cross for his daring under enemy fire during the crossing of the Senio. Abram Silberstein, the tall, handsome transport captain whose exploits as a British commando in North Africa before joining the Brigade had inspired incredible stories, many of them even true. Haim Laskov, a commander worshiped by the men in his battalion, but whom Peltz still saw as the skinny boy he had tutored in mathematics a lifetime ago at the Technion. Marcel Tobias, the aristocratic Viennese-born ladies' man—Where did he find all his women? Peltz was constantly wondering—court-martialed in Egypt for the beating he had given a British soldier who'd complained about the "dirty Jews." And there was Oly Givon, disconcertingly blond and blue-eyed, thin and sinister like an unsheathed dagger, his lips fixed in a perpetual grin, a veteran of the secret German Platoon. It was a collection of uncompromising men, murderously professional, and Peltz felt an immediate pride in having been asked into their company.

Finally, Carmi moved to the center of the room and spoke to the men. He did not attempt to put together an argument, a justification for the course they were setting out on. Vengeance was like faith. You believed, or you did not. You either accepted that blood would assuage blood, or you did not. The fact that the men had come to the farmhouse was proof of their commitment.

In his brusque voice, he told them how it would be arranged.

"This is family business," Carmi said toward the end of his brief remarks. "Only Jews can do what has to be done."

When Carmi was finished, he did not ask for questions. He was convinced there was nothing more to discuss.

Yet someone spoke up. We should set up a secret court, a soldier suggested. Put each Nazi on trial. Once the evidence had been presented and a man's guilt was definitively established, then—and only then—should they execute him.

"This is not the kibbutz," Carmi shot back. "There can be no debates. In this situation, the law does not apply."

"Who will be the judge?" the soldier argued. "Who decides whom we kill?"

"Me," said Carmi without hesitation.

"You?"

"I will be the judge and the jury."

When no one objected, the matter was considered settled.

* * *

Pinchuk refused to join the group.

"I will not. I cannot," he told Peltz the following day.

"Afraid, Arie?"

"Afraid?" Pinchuk repeated incredulously. "What about honor? Jewish honor."

"Precisely. Honor requires an eye for an eye."

"In cold blood? Without trials?"

"That was how the Nazis killed my parents. That was how they killed your parents. What will you tell your children, Arie? That you were in Europe, but you did nothing?"

"If I could kill the men who murdered my family, that would be something else. Another matter. But to go off and kill indiscriminately? I don't see the vengeance in that. Don't you understand? It's murder. It's wrong."

Peltz did not know how to respond. Pinchuk claimed there was a choice. But there was no choice. Only duty: to the memory of his parents; to his grandfather; to his people. Future generations needed to know that Jews had avenged the death of their brothers. "Family business," Carmi had called it, and Peltz agreed.

Peltz continued walking beside Pinchuk in the hills above Tarvisio, but he was filled with contempt for Pinchuk's squeamishness. He remembered when Gofton-Salmond had admonished him by saying, "Know what the problem is with you people? You think too much." Peltz looked at Pinchuk and saw what the colonel must have seen in him: a man who scurried to find excuses to justify a failure of courage. A man whose only scruple was his timidity.

For his part, Pinchuk was more astonished than angry. It seemed impossible that Peltz could genuinely believe there was any morality in such low, impulsive behavior. It was a corruption of their heritage. There was no possible honor in the victim's becoming the victimizer. This manner of vengeance, ruthless and primal, was an illusion. Pinchuk stripped away Peltz's high-sounding justifications, and all that was left was the sin of murder.

Pinchuk begged his friend to reconsider.

But Peltz was not even listening. Compromises were no longer possible. “The problem with Jews like you, Arie,” he said, “is you don’t know how to hate. I know how to hate.”

He turned and walked off alone.

About ten days later, a whispered message was passed to Peltz as he was eating lunch. He reported immediately to the farmhouse in Camperosso.

Carmi was waiting for him when he arrived. His friend was dressed in the uniform of a British military policeman.

“You’ve been promoted,” Carmi joked, handing him a British major’s uniform.

Peltz put it on. The pants were a bit short, but otherwise it seemed to fit. “How do I look?” he asked.

“Perfect.” Peltz turned toward the unfamiliar voice and saw Oly Givon standing in the room. He was dressed in a poorly cut brown striped suit, the vest pulled tight across his chest, and he wore his shirt collar buttoned but without a tie. His blond hair had been cropped quite close to his head, and it looked as if it had been cut in a hurry.

“Who are you supposed to be?” Peltz asked.

“I’ll explain on the way,” Carmi said. “We have a long ride.”

THIRTY-THREE

The narrow road curved steeply down from the Brenner Pass. On one side were mountains, large and looming in the twilight. On the other was only a thin gravel shoulder and beyond it a drop of several hundred feet. Yet Carmi drove as if oblivious to the danger, his foot steady on the accelerator. He was in a hurry. He was on his way to find a man, and then kill him.

Dressed in an MP's uniform, Carmi was at the wheel of a jeep that had the markings of the Royal Seventy-eighth Division on its fender. Peltz, a British major, sat in the back. Even in summer the air had a bite this high up, and Peltz's hands were shoved deep into the bellowed pockets of his tunic. Oly Givon, wearing his brown suit and a hat with a peaked visor, sat slouched beside him, out of sight of any passing vehicles. So far his caution had been unnecessary. The dark mountain road was deserted.

They had been driving for hours. When they left the farmhouse in Camperosso, Peltz had been talkative, animated by a pre-battle eagerness. But by the time they were waved casually through the military checkpoint at the Austrian border, Peltz's mood had dipped. There was something both bold and solemn in what they were setting out to do, and it began to weigh on him.

He noticed that Carmi, too, was silent, but this alone was not evidence of an uneasy mood. In the field, Peltz knew, Carmi assumed his operational demeanor. He did not talk, he gave commands.

Givon, however, had been relentless. All the way into Austria he had been telling stories about what it had been like in the woods of Mishmar Ha'emek training with the German unit.

On another evening, Peltz might have found the talk intriguing. According to Givon's string of stories, the men had initially been chosen because of their Aryan appearance and their fluency in German. Then they went through months of rigorous and inventive training to prepare them for one goal: to enable Jews to impersonate *Kameraden*, as the members of the SS referred to one another, in any situation, whether on parade, in the barracks, or in the beer hall. Awoken in the middle of the night, the only acceptable response to a command was not a simple *"Jawohl,"* but an unhesitant *"Zu Befehl"*—*At your command.* If a cagey instructor began ordering them about in Hebrew, they were expected to act as if the language was totally incomprehensible. Every aspect of Nazi lore—collar insignias for each rank, occasions when the various uniforms would be worn, political treatises, even marching and drinking songs—was studied and memorized. As proof, Givon serenaded them with several verses of the Horst Wessel song. And he had gone on singing with gusto until Peltz, exasperated, insisted that he stop.

Peltz knew that Givon was trying to reassure them, to let them know he was prepared for his role. But he was playing for laughs, trying to affect a humorous and ironic tone. More than the constant chatter, it was this lightheartedness that Peltz found both inappropriate and grating.

Givon ended the song, but he kept talking. He soon began a story about a man from his unit who had been nudged awake in the middle of the night by his girlfriend. "So of course he shouts, '*Zu befehl!*' And his girlfriend wants to know just who this Zu Befehl is . . ."

Before Givon could deliver the punchline, Carmi steered the jeep off the main road and down a rutted slope. The vehicle lurched to a stop in a stand of of trees. Below them, the ground sloped toward the banks of a fast-running stream.

Carmi shut off the headlights and turned to Givon. "You know what to do, Oly. Bring our man back this way. Just the two of you. We'll be waiting."

"Zu Befehl," Givon said. Then he jumped out of the jeep. Peltz watched as he headed quickly up the knoll, and disappeared into the night.

"Now what do we do?" Peltz asked.

"We wait," Carmi said.

In the first week the execution squads, the *huliyot*, had killed two men. The suspects were taken off for questioning and never returned. Yet when the execution teams next went out, they found their targets had fled. Carmi asked the American intelligence unit to make inquiries, but they uncovered no clues. The men, they informed him, had vanished.

Carmi's life in the underground had taught him that there were no coincidences. The targets must have been warned. Raging, he burst into the house of the Gestapo officer who had provided the list of names.

Carmi found the man in his study. Without a word of warning, Carmi drew his revolver and pressed it against the man's head. Then he spoke. "Tell me the truth," he said evenly. "Either you warned them, or somebody else did."

The man was trembling, but he insisted that no one informed the men on the list.

Carmi cocked the gun.

The man began to talk quickly. No one was warned, he repeated. But there was a network that gave SS men money and new identities so they could escape. It provided them passage on ships bound for South America. Perhaps that is what happened to the men you were looking for, the man said.

Carmi considered this. "I want a name. Tell me the name of the man in charge."

The Gestapo officer was sobbing, but he said he did not know any name.

Carmi once again pressed his revolver against the man's temple.

There's a beer hall, the man offered. It's their headquarters. He told Carmi the southern Austrian town where it was located.

"More," Carmi demanded. "Tell me everything. Who's in charge? Who runs the network?" With each question, he ground the barrel of the gun into the man's head with renewed force.

The man pleaded that he did not know. Finally, Carmi believed him and returned his weapon to his holster.

That night Carmi drove alone into Austria and found the beer hall. But he resisted the temptation to go in. He knew the risk was too great; he undoubtedly would give himself away. Instead he watched from a distance, a soldier on a reconnaissance mission.

He made a mental inventory of all the doors and windows. He studied the men who entered, how they acted, how they were dressed. He listened to the merry noise, the songs sung in loud drunken voices. In his mind he tried to picture the large smoky room full of happy, flushed-faced men: his enemies. He was still at his post hours later as he watched them stagger off to their homes.

Before returning to his jeep, Carmi walked with great attention through the neighborhood. First he toured the nearby park, looking for places where vehicles and men could be concealed. Then he walked along the stream, following the grassy bank as the water widened directly below a clump of pine trees, and went on to bend in a U around the beer hall's terrace. And finally he walked through the town's cobblestone streets, only to discover that with each step his military boots made a disturbingly loud tattoo in the quiet night.

On the drive back to Tarvisio, Carmi mulled it all over. He thought about how many men he would need. Where they should wait. And when they would make their move. Then he went to find Givon.

Givon listened, and without much reflection agreed that it might work.

Carmi did not involve Peltz until the afternoon of the mission. His part, the haughty British major, would require little preparation. But this was also Peltz's first execution, and Carmi suspected it would be easier if his friend did not have too much time to dwell on it.

In another life, Givon might have been an actor. He enjoyed playing roles, relishing the challenge of slipping into an assumed life. When he entered the beer hall that night, Oly Givon no longer existed; now he was Ernst Hacker.

According to the loose biography Givon had constructed, Hacker had served proudly as an SS lieutenant with the *Einsatzgruppen* as they swept through the Ukraine. But now the Allies were searching for him, vindictively hoping to put him on trial for war crimes. He had fled his home in Munich and come to this beer hall in Austria because he had been told—and he was prepared to use the names of the two SS men who had disappeared as references—that a soldier who had loyally served the Fatherland, a *Kamerad* down on his luck, could find help here.

But these were only the facts of Givon's new life. No less important was the attitude he projected. Ernst Hacker would be guarded, restrained by his own suspicions. He would instinctively repel intimacies that were too quickly offered. He would be proud, but he would also be very scared.

This was the Ernst Hacker who entered the beer hall. He found a place by the bar, and drifted in and out of conversations. He was aware that he was being appraised, measured in nonspecific ways by the men who spoke casually with him. But no direct questions were asked, and he was careful not to offer any information.

Soon, however, two men approached him. One was middle-aged and balding. He had an owl's face, round and fleshy, and he wore glasses. The other was tall, full of an officer's confidence and authority. It was this man who spoke. "Who are you?" he demanded. Givon immediately recognized he was a man who was used to issuing orders, and having them obeyed.

Yet Givon hesitated. Hacker would try to protect himself; he would be afraid.

"Perhaps we can help you," the man coaxed, trying to be reassuring. "But first we need to know who you are."

Givon told them all about Ernst Hacker, and the two men listened without interruption. He had assumed they would try to substantiate his story. He had been prepared to answer questions. But now their patience filled him with concern. Their silence, he was convinced, could only mean one thing: They knew he was an impostor.

As he talked, Givon worked out what he would do. He would pull out the Luger tucked into his belt by the small of his back and start firing. Perhaps he could make it into the street. Perhaps Carmi and Peltz would hear the gunfire and hurry over in the jeep. Against the eighty or ninety men in the room, perhaps the three of them could put up a valiant fight.

But the tall, authoritative man finally spoke. See that gentleman, he said, gesturing toward the back of the room. Go have a beer with him.

"Heil Hitler," the man said when Givon approached the table.

Givon clicked his heels and extended his right hand, palm down. "Heil Hitler," he saluted.

The interrogation began as soon as Givon sat down. Hacker was required to give an account of his entire military history, where he had served, the names of his commanders. He did this flawlessly.

From there, the talk moved on to Hacker's childhood in Munich. Hacker answered every question without hesitation. But this was to be expected; Givon had often visited Munich during the first seventeen years of his life.

"I think we'll be able to help you," the interrogator said at last. He rose from the table and told the lieutenant to follow him.

They walked across the room, and Givon knew he had met his target. This was the man who ran the operation. This was the man he would deliver to Carmi and Peltz. This was the man he would kill.

As they approached the door, however, the man steered him away, and into an adjacent room.

The owlish man with glasses was waiting. "He's a good friend, Colonel," the interrogator announced. Then he turned and left.

The bespectacled man gave Givon an appraising look, then spoke for the first time. "You must be hungry, *Kamerad*."

As closing time approached, Carmi grew anxious. From his hiding place in the trees, he could hear the men leaving the beer hall, their cheery voices carrying in the summer's night. But there was still no sign of Givon.

If they were keeping Oly inside, that could mean that they were taking him seriously. Or it could be that they had found him out.

"Johanan," he decided. "Let's get closer. In case Oly needs us."

"The jeep?" Peltz asked.

"Leave it here."

Carmi led the way down to the riverbank, and they followed it until they were hidden beneath the terrace of the beer hall. From this new position, they could stare up the grassy slope at the front door.

"What if they go out the back?" Peltz asked.

"They won't," Carmi said, praying he was right.

Givon had no appetite, but he knew Ernst Hacker would be famished. So he ate a huge plate of sausages and drank tankards of foamy beer. And he listened.

The colonel assured him everything would be arranged. The lieutenant would be given money, a new name, new papers. He would travel to Italy, and from there passage would be booked to South America. It would all happen quickly, perhaps even by tomorrow night. Until then, he would be a guest in the colonel's home. Is that satisfactory, Lieutenant?

"Zu befehl," said Hacker.

Their beers drained, the two men walked out into the quiet village street.

* * *

"There they are," whispered Peltz.

"See. The front door," Carmi said, while silently offering up thanks.

In spite of all the beer, Givon was alert. He had found his man. An SS colonel who had been stationed in Riga. Givon did not have to ask what the man had done in the war; he could imagine. Givon condemned him to death without a qualm. All Givon had to do was to lead the SS man toward the stream where Peltz and Carmi were waiting, and it would be done.

Come, the colonel told Hacker, as soon as they were outside. My house is this way. He started to walk up the cobblestone street.

"They're going the wrong way," Peltz excitedly whispered to Carmi.

"Wait," said Givon to the colonel. "I have to take a piss. All that beer."

Without giving the colonel any opportunity to object, he headed away from the street and toward the stream. The colonel followed. But Givon's mind was racing: *How am I going to get him to the jeep? We parked too far away. It was a mistake.*

Givon and the colonel were down by the water, both men about to urinate, when two armed British soldiers—an MP and a major—suddenly approached in the darkness.

"Ernst Hacker," announced the major in a crisp British voice. "You are wanted for questioning by the Allied command."

The colonel zipped his pants and started to walk off. He was trying to pretend this did not concern him.

"Sergeant," the major ordered the MP, "apprehend that man."

Carmi grabbed the colonel roughly and led him back to Peltz. "We will hold you for questioning," Peltz said. "If you are innocent, we will let you go."

The colonel started to protest; perhaps he did not even understand

English. But Peltz cut him off. "Sergeant," he said to Carmi, "take this man away!"

They led the colonel up the grassy incline and back toward the stands of trees. But it had been a mistake to park the jeep so far from the beer hall, and the colonel became suspicious.

"Who are you?" he demanded in German. "This is not right. What is going on?"

Carmi turned to Givon. "Now," he ordered.

Givon looked at Peltz. Without a word both men drew their revolvers and aimed them at the gaping colonel.

"We are Jews," Givon spoke in German.

"Don't shoot. Don't shoot," the colonel begged.

"In the name of the Jewish people," Givon said, "I sentence you to death."

He shot, aiming right between the man's eyes.

Peltz fired at the same instant.

The colonel went down as if his legs had been pulled out from under him. He lay on his back, his face shattered, a tangle of blood and tissue.

They carried his body down the slope and hurled it into the water. It landed with a splash, and the stream quickly carried it away.

On the way back to Tarvisio, Peltz found that his somber mood had lifted. He felt revitalized. "I know how to hate," he had told Pinchuk, and tonight he had proved it to himself.

Carmi was busy that summer. On his orders, execution teams went out nearly every night from the farmhouse in Camperosso. Each job was a military operation: covert, well-planned, disciplined. Often there were four separate squads operating at the same time throughout Western Europe, each *huliya* on its own mission, hunters on the trail of the hunted.

THIRTY-FOUR

Nighttime in a village near Mauthausen, Austria. Two men in British army uniforms, their red armbands identifying them as military policemen, pounded on the door of a small two-story house. At last a woman opened the door and let them in.

"Is this the home of Herr Weiner?" asked the officer.

The woman appeared confused. The stocky sergeant repeated the question in German.

"Yes," she said. "What do you want?"

"Herr Weiner is to accompany us to an interview at headquarters," said the officer.

"But this is a Russian zone," she argued in German. "The British have no right."

Just then a large ruddy-faced man hurried down the stairs. He wore his nightshirt over his trousers, and the soldiers wondered whether he had been in the process of getting dressed or undressed before he decided to investigate the commotion.

"Are you Herr Weiner?"

"Jawohl."

"You have been accused of war crimes," the officer announced in a proper English voice.

Before the sergeant could finish translating, the man began to protest. "This is the Russian zone," he said, echoing his wife. "You have no authority."

He turned to walk away. The sergeant grabbed him from behind and shoved him toward the door. Weiner stumbled to his knees.

He stood up slowly, and gave the two British MPs a thoughtful inspection. "You are impostors," he shouted angrily. "You want to kidnap me." Turning to his wife, he grumbled, "They're Jews."

Peltz looked at Carmi. This was not the plan, but there was little choice. Carmi gave a curt nod.

"In the name of the Jewish people," Peltz said, "I sentence you to death."

He took his Colt .45 from his holster and shot Weiner once in the head. Just to be certain, he fired again.

The woman was horrified. Peltz looked at her and she shook with terror.

"No," said Carmi. "Leave her."

Peltz returned his gun to its holster. When the two men walked out, they closed the door behind them.

Peltz learned that Zorea preferred not to use a gun. "Wet work," he called it dismissively. Not only was it messy, blood and skull fragments flying haphazardly about, but, Zorea explained, it was also dangerous: a stray bullet could hit one of your own men.

But Peltz had not been prepared for the alternative. In a forest in southern Germany, he watched Zorea put his thick hands around the neck of a concentration camp guard. Zorea squeezed with all his strength until the guard, gurgling, legs kicking wildly, finally collapsed to his knees. His eyes bulged from his crimson-colored face. And still the man would not die.

"Shoot the bastard," Peltz shouted. "Shoot him."

Zorea, face clenched, arms trembling from the exertion, continued to squeeze.

"Shoot the bastard," Peltz repeated.

Zorea's arms hurt, and he could not maintain his grip. When he let go, the man fell forward onto his stomach. But he was still not dead. Small gasping sounds came out of him. He lay facedown in the dirt, making indistinct noises, and Zorea put an end to it. He shot him in the back of the head.

"Kill them like you would kill vermin," Zorea told Peltz.

Givon, however, would never shoot anyone in the back. He needed to see the reaction when the victim realized that the man standing in front of him, preparing to execute him, was a Jew.

A report came in from the Pontebba Displaced Persons camp. A refugee recognized another woman from Ravensbrück. But she had been a guard, not a prisoner.

Carmi sent Givon, disguised as Ernst Hacker, to the camp to learn the truth. It did not take him long. When Hacker confided to the woman that he was an SS officer on the run, the suspect felt secure enough to share her secret, too.

The next day Givon took her for a walk in the hills around Camperosso. When they entered the abandoned farmhouse, Peltz and Carmi were waiting.

"In the name of the Jewish people," Peltz said, "I sentence you to death."

But this time he was sweating beneath his uniform when he shot her in the head.

Zorea was determined to find a more efficient way of strangulation. After many attempts, he finally succeeded.

They would put the prisoner in the back of a truck, a guard seated on either side of him. Two other soldiers would be facing them.

A few short jams on the brakes would be the signal to start the operation. Then one of the men across from the prisoner would take out a pack of cigarettes and considerately offer him one. The soldier would light his own cigarette, then the other man's. When the prisoner exhaled the first puff of smoke, the two guards sitting beside him would swiftly wrap a rope around his neck, and pull with all their might.

"That's how you strangle," Zorea explained. "And it works well."

Carmi kept no record of the targets that were eliminated that summer. The risk of an Allied tribunal finding this document and using it as evidence was too much of a danger. The men would be court-martialed, but the political consequences for the Yishuv would be disastrous.

The men, too, had quickly stopped counting their victims. They estimated that at least two hundred people were executed. Perhaps, some of the participants insisted, the total was closer to three hundred. But they were not gunfighters carving notches into the handles of their Colts. The only number that mattered was the number of Jews who had been killed by the Nazis. So they did not bother to count. And they could not stop.

Peltz found it was always with him. Even when he tried to get away, he could not escape. The responsibility to hunt down the murderers was a constant presence: one act of vengeance cycling inevitably into the next.

One afternoon as he hiked by himself in the foothills of Mount Accmitza, he saw a cabin in a remote meadow. He felt he had to have a look. He crawled toward the window and peeked through.

There were fifteen people inside, men and women. He was certain he had stumbled on a Nazi hideout. Perhaps one of them might even be Joachim von Ribbentrop. It was possible, he told himself. The Allies were looking everywhere for Ribbentrop, but leave it to Johanan Peltz to find him. In the mood he was caught up in, it made perfect sense.

He raced back to the camp and assembled six men from the elimination squads. "This is the big one," he told them. The team returned to the cabin.

They broke in at dawn and herded the fifteen people into a line at gunpoint. Von Ribbentrop was not one of them, but Peltz refused to give up the hope that he had found a nest of Nazis. He ordered the men to remove their shirts and raise their arms. Two had blood groups tattooed under their left arms.

Peltz instructed that the other thirteen prisoners be led back to camp and delivered to the intelligence unit.

He and Givon took charge of the two SS men. They marched their prisoners up a trail that led toward the mountain's summit. After a while, he instructed the SS men to stop. Training his rifle on them, he ordered the men to take one step back. Then another. And another. Until their heels were near the edge of a rocky precipice.

"In the name of the Jewish people," he said, "I sentence you to death."

Givon pushed one man, then the other, over the ledge. Peltz stood there, fixed to the spot, as he heard two voices trail off in a plaintive, receding echo.

"I know how to hate," Peltz had proclaimed. But as his reckless summer rushed on, he was discovering what was required to nurture and sustain this kind of rage. And it was becoming more and more difficult.

THIRTY-FIVE

Peltz made a decision that summer. There had been no internal debate, no period of reflection. He had reached it suddenly. It came to him, and he accepted it. It was his secret, and he shared it with no one.

But once it was in his mind, he found confirmation wherever he looked. He would walk through the Brigade's camp in Tarvisio, see the troops, and would think how naive he had been. There was no possibility of brotherhood with these men. He could not share their world. He was different.

He had left the ruins of Zabiec, and in his sorrow he had entered into a new sort of existence. He had become an executioner.

Each time he pulled the trigger of his heavy Colt, each time he watched the head of one of his victims explode, the intensity of his rage increased. And the ordinary world receded, moving farther out of sight, beyond the horizon. He knew how to hate, and this was his new homeland.

The procession of killings, murder after murder after murder, had changed him. A normal life was no longer possible. Vengeance was all that remained for him, and he was devoted to it.

And this was his secret: He was not going back to Palestine. He would stay in Europe. He would avenge his parents, his grandfather, the memory of everything that should have been his but had been taken away. He would kill Nazis for as long as he could. Or, as he increasingly suspected, until they killed him.

Guarding his secret, Peltz deliberately cut himself off from his previous life. He ignored his battalion duties. He ate alone. He no longer had any use for Pinchuk. And he barely spoke to Carmi. All he needed was the location of the next target.

But when Abba Kovner, the leader of a Jewish partisan group that had operated behind the German lines during the war, arrived in Tarvisio, Peltz was eager to hear him speak. He admired Jews like Kovner, men and women who had stood up against the Nazis. It was the way he would have acted.

At a secure meeting in the farmhouse in Camperosso, Kovner told the soldiers in the elimination units about his new plan. He wanted to poison the water supplies in cities throughout Germany. Munich, Hamburg, and Nuremberg would be his first targets. "They poisoned us. We will poison them," he declared. He was determined to kill as many Germans as possible. Men, women, children—it did not matter. Everyone was guilty. His ambition was to claim "six million for six million."

He had come to Tarvisio to ask for the Brigade's help. He wanted the soldiers to use their sources in Palestine to obtain the lethal poison. And he hoped to recruit them to participate in the actual operations, casting the poison into reservoirs and, if necessary, storming water pumping stations.

When Kovner finished, the men looked to Carmi. It was his decision. They would follow his orders.

Carmi understood Kovner's determination. And he was intrigued: Revenge on an entire nation whose official policy was the annihilation of the Jewish people would make history. It would be a response that would make future anti-Semites tremble.

But, ultimately, Carmi could not support such a plan. The killing was too indiscriminate. There was no possibility of honor in mass murder.

He felt no guilt for executing individuals who had participated in the war against the Jews. He was a soldier and he could justify taking revenge on his enemies. But he could not make the moral compromises necessary to participate in Kovner's far-reaching scheme.

"I am sorry," he told Kovner. "You have your way, and we have ours. We use tweezers. You want to use a bulldozer."

Peltz was bewildered, and angry.

How could Carmi decide that killing two hunderd Germans was correct, and then argue that killing two million—*two million!* he told himself with excitement—was wrong? Either avenging the dead was moral, or it was not. There could be only one reason for Carmi's refusing to join in Kovner's visionary plan. And after Kovner had left, Peltz could no longer hold back.

"You're afraid," Peltz accused. He was standing outside the farmhouse in a group of lingering soldiers, but his words were clearly directed at Carmi.

"I think we should let this go," Carmi said evenly.

"Shooting men, women—that's easy. It takes guts to go into a German city."

"Guts?" Carmi repeated, astonished.

"You're afraid, Carmi."

"Let it go."

"Like hell I will. Coward."

"Screw you."

Peltz rushed Carmi, his fist raised. But before he could strike, Carmi grabbed the taller man around the waist and threw him to the ground. The two men rolled in the dirt, struggling for an advantage, until several astonished soldiers pulled them apart.

Peltz walked off, refusing even to look back at Carmi. Later that night, alone in his room but still tormented, he found himself wondering, *What is happening to me?*

A colossal, seething anger had taken control of his life. He was helpless against it, and he suspected he was going mad. But, he finally came to decide as the long, restless night dragged on, if he was losing his mind, it should not bother him. It would only make the serious work ahead that much easier.

Carmi understood what had pushed Johanan out of control. With each execution, it grew more difficult to pretend that a bullet fired at point-blank range had no effect on the shooter. He was their commander, but there were moments when his orders made him uneasy, too.

When Carmi learned that British intelligence was preparing to arrest the couple who had provided the original list of names, his initial reaction was that it was none of his concern. He had the information he needed.

Besides, the man and woman were evil. They had helped deport thousands of Jews. They had stolen Jewish property. They deserved to be arrested.

Then he pictured their interrogation. How long would it be before that witch disclosed that her husband had provided soldiers from the Jewish Brigade with the names of high-ranking Nazis?

And with this image in his mind, Carmi realized he could not let them be arrested. He would have to break his promise. Carmi had given them his word: if they cooperated, they could live. But now the chance that the British would blunder into the clandestine vengeance operation was too great.

Carmi arrived at the couple's house without warning, explaining he needed help in sorting through a chest crammed with Wehrmacht documents his men had discovered. Immediately, both the husband and wife were suspicious. They were reluctant to leave; Carmi had to lead them to his jeep at gunpoint. And it could not have been reassuring when the British MPs manning the roadblock by the forest waved Carmi through without a word.

Carmi drove deep into the woods, and came to a halt in a glen of towering conifers. It was a world of complete silence, a place for secrets. When he pointed toward the cave where the chest was hidden, he could tell they did not believe him. The couple knew they were doomed. They were sobbing as they walked ahead of Carmi, their feet slowly trudging through the pine needles.

When they got closer to the cave, he ordered them to stop. "Turn around," he said.

The couple turned to face him.

"In the name of the Jewish people," he said, "I sentence you to death."

Carmi shot the man, and then the woman. He did it rapidly, without pause. "The minute you start hesitating," he often told his men, "you might as well go home." Carmi never hesitated when he executed people.

But for the first time he found he wanted to.

In the middle of July, Carmi asked Peltz to accompany him on a mission into Poland. He chose Peltz because it would be dangerous, and Peltz was someone he could count on. He also wanted Peltz to understand that there was no lingering bitterness on his part. The words that had been spoken were meaningless. Their fight was a mutual childishness.

But there was still another reason why he selected Peltz. He felt a growing sense of kinship. He would look at Peltz, at the detached taciturn presence he had become, and he sensed they both were converts to the same resigned faith. For Carmi now had a secret, too: he no longer cared what happened to him.

THIRTY-SIX

During the long trip across Europe, the two men hardly talked. There was no residual anger, but both Peltz and Carmi felt there was little point in conversation. On this mission, their only connection was operational. The experiences they had shared belonged to other men. It was as if the two friends who had ridden with the Palestinian Police, challenged one another to weather out the storm on the boat to Italy, whistled at the girls in Rome, no longer existed.

And they were being cautious. If they remained silent, perhaps the anarchy that raged inside them would go undetected.

The job itself was direct enough. Carmi had learned through American intelligence about an SS officer hiding in a church. They would find him, and then execute him.

There had been a time when an action would only have been launched after much planning. But there had been no reconnaissance of the church, no mapping of an escape route through the town. Military precision no longer seemed important. Carmi matter-of-factly had told Peltz that they would work things out in the field. With similar unconcern, Peltz had agreed.

The church was located in a small medieval town in Trans-Olza. Before the war the region had belonged to Czechoslovakia.

According to the map British intelligence had given them, the area was now considered to be part of Poland. It made no difference to Carmi and Peltz. They had a target and they were prepared to hunt him down.

They drove on.

About ten miles north of the town, they decided to stop. They would camp in a forest for the night. In the morning, the two MPs would enter the church.

Carmi, exhausted from driving, woke up late the next day. He shaved and rinsed his face with water from his canteen. Then he saw Peltz.

Peltz was wearing the uniform of an officer in the Jewish Brigade.

"What are you doing?" he asked, staring with astonishment at his friend.

With great calm, Peltz explained that he no longer wanted to be an impostor. He wanted to enter the church as the man he was: a Jewish soldier.

"You think they'll let two Jews drag a priest from a church?" Carmi argued. "They'll get the police. They'll try to stop us."

Peltz, his Colt holstered at his hip, his rifle slung over his shoulder, said, "Let them try."

When Carmi joined Peltz in the jeep, he wore his sergeant's uniform. If Peltz was wearing the Brigade flashes, there was no point in his being disguised. He would also wear the uniform of a Jewish soldier. And if the Poles, the Czechs, the police, or even the priests dared to stop him, he was willing to let them try, too.

They drove into the tiny town, and located the church with little problem. Its spire rose above the other roofs, and using it as their guide, they headed down a series of sloping streets until they saw the large, vaguely Gothic structure. They parked the jeep opposite the church's front steps just as it started to rain.

The two men crossed the stone threshold and entered the cool, dark interior of the church. Peltz had his rifle strapped over his shoulder. Carmi had his revolver in the holster on his hip.

They walked up the main aisle. Votive candles flickered in the faltering light. A sweet, aromatic fragrance lingered in the air. They continued down the rows of pews, their footsteps tolling on the ancient stone.

But then they stopped. As they made their way across the nave, first Carmi, then Peltz, came to a halt.

A group of girls, perhaps as many as a dozen, were seated in the bay across from the altar. A black-robed priest was standing in front of them. Next to him, as though assisting in some ecclesiastical task, was a nun.

Ignorant of church matters, Peltz and Carmi assumed they had walked into a service, and they came to a halt.

But they were not simply being respectful. The sight of the young girls penetrated their mood. After the drastic, murderous way they had been living their lives, they were unprepared for such normalcy. The girls were fresh-faced, fair-haired, pigtailed. They were youth and innocence; and their existence seemed a rebuke to all that the two men had become.

Carmi gestured to Peltz and the two men sat down. Their unspoken plan was to allow the service to conclude, and then they would proceed with their mission.

The priest and the nun avoided eye contact with the two armed soldiers, but the girls did not try to control their curiosity. From time to time, they would turn and stare candidly at the two men.

Soon one of the girls got up from her seat. With her dark pigtails bouncing, she hurried over to Carmi. Yet as she got closer, she hesitated. It was only when Carmi smiled at her reassuringly that she continued forward.

She went up to him and pointed to the yellow Star of David on his shoulder flash. "Magen David," she said. She seemed eager to show off her knowledge.

"Are you Jewish?" Carmi asked in Yiddish.

"Yes," she said.

The nun was calling to her, but the girl ignored her.

"Where are your parents?" Carmi asked.

The priest started to approach. Peltz quickly got to his feet. He moved into the aisle and stood in front of the girl. The priest looked at the tall, fierce soldier and retreated without a word.

The girl told Carmi her story. Her parents had been killed in the camps. Her aunts and uncles were gone, too. She was an orphan, and the nuns had taken her in. She spoke without emotion, as though she were talking about someone else.

"Do you want to stay with the nuns?" Carmi asked.

"No," said the girl emphatically. "They say I have to be Catholic. I am not a Catholic."

"Where do you want to go?"

"Where there are Jews," she answered.

Carmi held her response in his mind. He weighed it, and the consequences.

"Johanan," he said to Peltz in Hebrew. "This girl is Jewish and she wants to go to her people."

"Then we take her," said Peltz at once.

"They won't like it."

"Good," said Peltz.

Carmi offered the girl his hand and she grabbed it. He led her down the aisle, past the candles glowing in the narthex, and toward the door. Peltz walked slowly behind them, his rifle cradled in his arms, facing the others.

The nun yelled at them. A few of the girls called, too. The priest said nothing. He simply stared at the tall soldier, the eyes in his handsome boyish face filled with hate.

Peltz remained in the doorway, his back to the street, on guard, as Carmi helped the girl into the jeep. When the engine was running and Peltz was convinced no one was coming after them, he turned and walked slowly down the steps.

* * *

The girl's name was Eve. She was Polish. She was twelve years old. They had rescued her from the nuns, but now the two soldiers had no idea where to take her. All they knew was that she wanted to be with Jews.

Carmi suggested they find a Jewish family in Poland. But Peltz, knowing how the Poles had treated his family, would not allow it. Carmi offered another idea. He had read in the Intelligence reports that thousands of Jewish children were being temporarily sheltered at Bergen-Belsen, a refugee camp that had been established on the site of the former concentration camp. But as soon as he tried it out on Peltz, he realized it would be unforgivable to leave Eve in such a hopeless place. When they had run out of alternatives, they decided to take her to the transit camp across the Italian border at Pontebba. It was near the Brigade's base in Tarvisio; at least they would be able to look in on her from time to time.

On the ride to Pontebba, Carmi told Eve about his young daughter who lived on a farm in a land where melons and oranges grew in the warm sun and it was not a crime to be a Jew. She told him it sounded like he lived in Heaven.

Carmi left Eve in a barracks crowded with other Jewish children. "We'll meet again," he promised her.

Returning to his jeep through the camp, Carmi passed a group of survivors playing Mozart. A relief worker had rounded up a collection of musical instruments, and they were having an impromptu recital. Carmi listened, and for the first time in months he was able to find refuge in a calmer place.

When the piece was finished, Carmi approached the teenage cello player and praised his performance. The boy waved off the compliment. He gripped Carmi tightly by the arm and said, "When we get to Palestine, I'll throw away the cello and you'll teach me how to use a gun. Yes?"

That night in his room in Tarvisio, Carmi thought about the boy's words. And about Eve. And about the other Jewish children, orphans on their own, in Europe. And in these heartfelt thoughts, in these affecting images, he began to recognize the promise of something new. In them he saw a way for him—for Peltz, for the *huliyot*—to transcend the hopeless state they had locked themselves into over the long reckless summer.

He had told Peltz that they would return to the church for the SS man. But his newfound clarity had revealed the futility of his violent campaign. He could not win. He could go on killing, but he could never bring a single dead Jew back to life. The acts of vengeance would only perpetuate a continuing pattern of murder. The more he killed in cold blood, the more he ensured that the horror the Nazis had let loose would continue to triumph. His only hope was to make a movement away from this ruinous faith. And now he knew what he had to do. For the first time he started to envision the beginnings of a plan, an active strategy, that brought with it the possibility of a world beyond all the evil.

It was not long after Carmi had worked this out in his mind that Capt. Pinchuk's batman came into his room with the mail. Pinchuk rarely got letters, and when he did they were always inconsequential. He assumed the letter that was handed to him, forwarded from his flat in Tel Aviv, was simply another nagging request from the university. Last month the dean, with an unworldliness that struck him as typical of the academic mind, had written to say that unless Mr. Arie Pinchuk promptly notified the university of his intentions, they could not guarantee a place when classes resumed in the fall. When Pinchuk opened the envelope, however, he found a telegram from Moscow.

He read it once and, still unable to believe the words, tried reading it again. But he could not finish. His eyes were flooded with tears of pure joy.

PART V

EUROPE

Summer–Spring 1945–46

THIRTY-SEVEN

Carmi arrived in Paris early on a hot July morning. There was the sight of pavements being washed down with water, and the smell of fresh baked bread from the *boulangeries*. Couples sat at the cafés leisurely reading the papers; by their dress Carmi suspected they were still on their way home from the night before. Bicycles and big green-and-white municipal buses swooped down the boulevards. The city had shrugged off the war. He was intrigued, but he had no time for diversions.

The apartment was just off the Champs Elysées, on the second floor of 53 rue de Ponthieu. Carmi knocked, and the door was opened by a man whose face was on Wanted posters throughout Palestine.

"*Shalom,* Yehuda," said Carmi warmly.

He had not seen Yehuda Arazi since they had planned the theft of the crates of Brens from the British base in the desert on Christmas Eve. After another *rechesh* operation, the Mandate authorities had tried to arrest Arazi; at the last moment he had noticed the police hiding outside his house and fled. Disguised as a sergeant in the engineering corps, he had ridden in a troop train to Egypt. From there he had hitched rides across North Africa to Tripoli. Impersonating a Polish airman, he flew to Rome in a British bomber. And now employing a dozen different aliases, he had become head of the Haganah Field Services for Italy.

They shook hands, but both men knew now was not the time to reminisce. Arazi swiftly led Carmi into a sunny kitchen. Seated at a table covered with a white cloth was Nahum Shadmi, a glass of tea in front of him.

Shadmi was an imposing gray-haired man. Dressed in a dark suit and a striped tie, he looked like a diplomat. But Shadmi was a soldier. In World War I he had been a Russian officer. After immigrating to Palestine in 1921, he found his way into the underground. He was serving as commander of the Jerusalem District of the Haganah when Ben-Gurion ordered him to Paris. Despite the stiff, formal image he presented, Carmi knew from experience that Shadmi had been a shrewd choice to direct the Yishuv's secret activities in post-war Europe. He could be discreet, inventive, and, if the situation required, utterly ruthless.

Carmi was eager to explain why he had requested this meeting. He had been formalizing the idea since the day he had brought the young girl out of the church in Poland. He had inspected it from every angle. He was convinced he had devised a way of making a crucial contribution to the Yishuv.

At the same time—and this was the part he was not prepared to share—it was also a personal strategy. It was a way for both Carmi and his men to escape from the spiral of demoralizing violence that had come to dominate their lives.

"Nahum," Carmi said as soon as he sat down, "the Brigade can make sure that every boy or girl who wants to grow up in Palestine will have their chance. You find the ships," he challenged. "We'll bring the people to fill them."

When Carmi read the intelligence reports that had been prepared for the Nuremberg trials, he had been left numb, bewildered, by the numbers: 5,993,900 Jews dead; seven out of every ten people that made up the Jewish population before the war no longer existed.

The Nazis had set out to annihilate European Jewry and they had come horrifyingly close.

But they had not succeeded. It was a twelve-year-old girl named Eve who wanted to be with her people and a dark-haired cello player who wanted to be a Jewish soldier that made Carmi fully appreciate that there were survivors. And that he had a duty to them, too.

In all of Western Europe, perhaps 750,000 Jews—there was no census—had survived the war. About a quarter of a million of them, mostly concentration camp survivors, were sheltered in more than ninety refugee camps in Germany and Austria. It was a pitiable existence.

In June, a delegation of five soldiers from the Brigade had toured the camps. When they returned, Carmi found a typewriter so that Capt. Hoter-Yishai could send his angry report to both the Allied command and the Yishuv.

Special Report About the Situation of Jewish Refugees in "Displaced Persons Centers" in Austria and Germany

. . . There are two types of Centers, mixed camps and camps which hold only Jewish refugees.

The mixed camps, e.g. Schleissheim, Garmisch, etc., Jews and non-Jewish refugees are herded together, a fact which caused clashes in various camps because of the antisemitic attitude on the part of the non-Jewish inmates of the camp. . . .

There are three grave problems to be dealt with. The first is insufficient supply of food and lack of clothing. The people who live in these camps were for years kept on a starvation diet in concentration camps and are now hungry, and undernourished. The fact that they have been put on the same ration-level as the civilian population,

> which has still food reserves, causes much bitterness . . . and they do not enjoy a better treatment than our former enemies.
>
> The second is a psychological problem. . . . Everyone is most anxious to know whether any member of his family is still alive, but there is nobody whom they could ask for information.
>
> The third problem is the question of repatriation and immigration. Many Jewish refugees both from allied and enemy countries are still afraid that they will be forced to return to the country they had come from. . . .

Hoter-Yishai managed to present a copy to the American special envoy to the refugee camps in his Frankfurt office, and conditions slowly began to improve. Food allotments increased. New clothes replaced the black and white prison outfits. And for their own safety, Jews were segregated from the other refugees. After the orchestrated daylight pogroms in Poland and the Ukraine, the policy of sending Jews back to their native countries was also rescinded.

But there was no reconsideration of the immigration policy.

On June 18, 1945, the Jewish Agency had optimistically applied to the Mandate government in Palestine for 100,000 immigration permits for survivors. There was no response.

The Yishuv waited as Jews in the refugee camps continued their grim existence. In Bergen-Belsen alone in the month following liberation, 13,000 more Jews died.

They waited as in Kielce, Poland, where Peltz's father had run a hospital, returning Jews were hacked with axes in the streets while policemen watched. Days later, the Polish government issued an astonishing edict: All Jews, including the estimated 120,000 who had returned from the Soviet Union, were to remain in Poland.

They waited as the British government closed down the borders of its European zones, keeping the refugees out of France and

Italy and away from the southern ports. President Truman urged the prime minister to grant the 100,000 permits the Agency had requested. Yet ironically, only 12,000 Jewish refugees were allowed to immigrate to the United States in the three years after the war.

They waited as Clement Attlee and Labour won the British general election that summer and Churchill's wartime government was dismissed. The Colonial Office informed the Jewish Agency that only 1,500 immigration permits would be issued—and these would be the last under the provisions of the White Paper. The new British foreign secretary, Ernest Bevin, in his first official statement on the refugee problem, scolded, "Jews should not push to the head of the queue."

They waited as, Chaim Weizmann forlornly observed, "The world was divided into two parts. One where Jews cannot live. And another where they cannot go."

The survivors, however, could not wait. In defiance of governments and laws, oblivious to reports and decrees, they acted. The Escape—the *Bricha*, in Hebrew—began.

These Jews had not made it through one war only to succumb to the peace. The desire to live as they had once lived, as families of husbands and wives and children, diminished all obstacles. From every corner of Europe, the Jews traveled west. They hiked through dense forests, trudged through snowy mountain passes, rode in inhumanly crowded trains. They were determined to make their way closer and closer to the coast. And to a ship that would take them illegally to Palestine.

At first the Bricha was spontaneous; these were the bewildered refugees who had drifted into the Brigade camp in Tarvisio. But the Haganah's *shlihim*—the emissaries sent secretly from Palestine, men like Shadmi and Arazi—had moved quickly to set up an underground network to direct the Jews in their flight across Europe.

Still, it was a makeshift operation, a handful of Palestinians trying to outsmart armies and governments. Thousands of refugees remained trapped in camps, detained at the borders by British soldiers, prevented from boarding ships that would take them to new lives.

The men from the Brigade had in countless ways already assisted the refugees they met in Italy. They had offered up bundles of gifts, from candy bars and loaves of bread to shoes and clothes. But, however much these tokens were appreciated, they were only small tendernesses.

Carmi had come to the apartment on rue de Ponthieu to offer something much grander. A product of both the military and the underground, he understood what the Brigade could do. Not only were his men positioned near three borders, but his *huliyot* in their months of terror had also learned the trick of slipping through checkpoints with impunity. He had a well-trained organization already in place, and it had a treasure of resources—food, trucks, clothing. They could move thousands of people across Europe to the Mediterranean ports.

The Brigade would run the Bricha as if it were a military operation. They would be a Jewish army rescuing their brothers. And at the same time, Carmi would make his own escape—from resignation to his own salvation.

There was no debate. Shadmi eagerly endorsed the plan, and Carmi left Paris that night. On the way back to Tarvisio, he thought a good deal about the newest irony that was shaping his complicated life. He had come to Europe as a soldier with a soldier's mission: to kill. But now he was about to embark on another task: to save lives. And it would be no less dangerous.

THIRTY-EIGHT

Where was Leah?

That was the question Pinchuk asked as he reread the telegram from Moscow. He was still filled with an extraordinary happiness. That his sister was alive was miraculous; even as he had given up hope, her telegram had been making its way back and forth across oceans and continents. But with each new reading, he grew more practical.

The information he had received was frustratingly brief, almost cryptic. The entire message was just two short sentences. Leah had survived the war. She would make her way to Palestine.

But where was she now? His sister could be almost anywhere in Europe. Borders had been closed, but chaos and ingenuity remained the ruling authorities. Was she trapped in Poland? Or had she managed to escape? Perhaps she was in a DP camp. But which one? In what zone? Or, no less improbable than anything else, she might at this moment be in Italy. For all he knew, Leah could be just miles away. There were so many possibilities, each entangled in its own complications.

But it did not matter. Now that Pinchuk was certain Leah was alive he felt all his previous decisions had been vindicated. He had been right to join the Brigade, to come to Europe.

When he had learned the dimensions of the Holocaust, he had despaired. In his grief, he had foolishly given up hope. Now he was being offered another chance. The telegram was a summons to fulfill the mission he had originally embarked on, and Pinchuk would obey.

Wherever Leah was, he would find her. And he would deliver her to Palestine.

His plan was quickly formulated, and wishfully simple. Starting from Italy he would work his way north, retracing the refugee routes, but in reverse. In that way the farther Leah had already traveled, the sooner he would find her. Even so, the search could take weeks, months, possibly longer. There was no way he could continue to fulfill his daily responsibilities as an officer in His Majesty's Army. He would have to desert the Brigade.

His mind was made up, and yet as he worked out his plan he began to feel guilty about betraying his commission. His sense of duty left him troubled. And while his priorities did not waver, he wondered if he should inform Col. Gofton-Salmond.

The risk was that the colonel would try to prevent him from leaving. But, Pinchuk reasoned, short of the colonel's throwing him in the stockade there was no feasible way he could be stopped. At least he would have done the honorable thing. He would have informed his superior officer of the reason for his disappearance.

There was also another, more personal allegiance that was pushing Pinchuk to make a confession. He had worked and served closely with the colonel for nearly two years. Even if the colonel's rank prevented him from condoning it, Arie wanted to believe that Gofton-Salmond would appreciate the higher duty that compelled him to find his sister. He wanted to believe that the "old man," although not a Jew, could still be his friend.

So Pinchuk marched into the colonel's office and announced that he was preparing to leave camp. After an angry exchange, the colonel agreed to listen to the entire story; and soon the old man's face began to soften.

When Pinchuk finished, the colonel insisted there was no need for him to go absent without leave. Instead, he would send an urgent personal telegram to the headquarters of the British Army of the Rhine. He would request that Capt. Pinchuk be given official authorization to search for his sister.

Arie was stunned. This spontaneous act of kindness, he decided, was a harbinger, a confirmation that his mission would be successful. A grateful Pinchuk assured the colonel that he would remain in camp until he received Command's approval.

Two days later a runner from battalion headquarters informed Pinchuk that the colonel wanted to see him on the double. Pinchuk hurried to the office and, without a word, Gofton-Salmond handed him a telegram. He read:

> Capt. A. Pinchuk will NOT repeat NOT proceed to Eastern Europe.

Pinchuk was so shaken he was unable to speak. He had tried to act honorably, but now he realized he had been naive. He would still go, but it was suddenly an escape, and it would require much more daring. He felt betrayed, and he hoped he could find the cunning to deal with these new circumstances.

But for the moment all he could do was look beseechingly at the colonel. He could not even find the words to express his disappointment.

Then the colonel broke the silence. "When will you be ready to leave?" he asked.

Suspicious, Pinchuk took a while to answer. "In a week, sir," he managed at last.

The colonel gave him a sly, conspiratorial smile. And he asked, "Why not tomorrow?"

A kind-hearted sergeant major from a Canadian battalion camped near the Brigade provided the jeep. The Canadians were going home, the sergeant major told Pinchuk. Shouldn't take much magic, he added with a wink, to make one vehicle disappear.

Pinchuk found a trailer in the Brigade motor pool and attached it to his new jeep. Uncertain of how long he would be away, he packed the trailer to its top with provisions: jerry cans of petrol, cases of field rations, spare parts for the jeep, two clean uniforms, cartons of the Players Light that he smoked, tins of Craver A cigarettes that were as sound as any currency in postwar Europe, and, even better, bottles of whiskey and gin in case he needed to ransom his way out of a genuine crisis. He also took clothes for Leah: military trousers, a skirt, and a cap. He guessed at the sizes. The last time he had seen Leah she was a skinny ten-year-old; now she was a young woman. Would he even recognize her? he worried.

The colonel had arranged for Pinchuk to receive a month's pay in advance. Pinchuk stuffed the currency, a thick wad of pound notes, into a khaki army pouch, and then tied it to a belt he would wear concealed under his uniform blouse. When he tried it on, there was only a small bulge, like a telltale layer of fat, and he was pleased with his invention.

Then he went to the officers' mess to have lunch. He was anxious, a man without much appetite, but he forced himself to eat. He did not know when or where he would get his next hot meal. He did not know when he would see these men again. The colonel stared at him knowingly, but nothing was said. They both wanted to protect their secret.

As he was eating, a group of officers invited Pinchuk to join them later for a rubber of bridge. He said he would meet them in the officers' club.

Instead, he finished his meal and went to his jeep. He checked that the trailer was firmly attached, climbed behind the wheel, and drove off.

It was only after he had been driving for a while, as he was reaching into the glove box for a pack of Players, that he saw the envelope. He pulled over to the side of the road and opened it. Inside was a single piece of paper.

"Good luck, Arie," it said. It was signed, "Gofton-Salmond."

* * *

The camps in Italy were a sea of faces, and he studied each one, hoping for the moment of sudden recognition. But there were only blond girls who might have been his sister. There were only refugees whose singsong Yiddish reminded him of his parents. There was no one he knew. He saw only his fellow Jews, and it broke Arie's heart to see what had become of them.

He had a map with red dots marking the locations of each refugee camp in Germany and Austria. On the night before he left Italy he counted the dots. There were ninety-two. He was determined to visit them all. From Mestre, near Venice, he drove north.

Each time he arrived in a camp—Feldafing, Allach, Bad Reichenhall—there were moments of high drama. People rushed up to him, desperate and bewildered. Was it possible there was such a thing as this young man on God's earth? A Jewish soldier, an officer with three captain's stars on his epaulets, a Star of David on his shoulder flash, a holster around his waist? Time after time, Pinchuk had to convince them that he was real, flesh like their flesh, a Jew like them.

These encounters affected him, too. He came to realize that he was not only a brother looking for his sister. He also had a public mission.

Now when the refugees gathered around him—sometimes he climbed onto a stage, sometimes he simply stood outside a tent—he did not simply ask in Yiddish if anyone had seen Leah Pinchuk of Reflovka. He also told them about Palestine. He wanted them to know that they had a home. He wanted them to understand that there were people waiting for them, waiting to welcome them. He begged these downtrodden, squalid congregations not to lose hope. *"Am Yisrael chai,"* he told them, repeating the ancient benediction, its message nothing less than valiant and defiant in such grotesque circumstances. "The people of Israel live."

He would cry, and they would cry. And then he would move on, eager to continue his journey before his own hope was lost.

* * *

The accident happened outside Mannheim. It was a little after six in the morning, the sun was rising in the sky, and the jeep was speeding along the Autobahn. As he steered into a curve, the trailer turned over and the jeep skidded down an embankment, into the forest, and smashed against a tree.

The Americans, who found Pinchuk and kept him in a hospital for five days, told him that if he had not been wearing both his battle dress and coat he would have certainly broken his shoulder. They also told him it was not an accident. The Werewolves, young unrepentant Nazis, often spilled hot oil on the Autobahn to sabotage Allied vehicles. But here, too, they explained, Pinchuk was fortunate. Sometimes the Werewolves stretched steel wire across the road: a man in a jeep would be beheaded as he drove by.

The trailer was dented, but when the Americans put on another set of tires it seemed stable enough. The jeep was a total loss. After the Americans heard Pinchuk's story, though, they told him not to worry. They gave him a new jeep, and six days after the crash he was back on the road.

Pinchuk visited more camps in Germany, and then he moved on to Austria. As the weeks passed, the news about a Jewish soldier from Palestine who was looking for his sister often preceded his arrival. The refugees were expecting him. And so were the officials who ran the camps. One indignant British officer had already warned Pinchuk that he was reporting him for encouraging the Jews to immigrate illegally to Palestine. So Pinchuk grew wary when, while standing in the midst of a crowd of people in a camp in Austria, he heard the hissed words, "Here comes the goy."

The refugees stepped back as a tall British major marched toward Pinchuk. "Captain," he ordered brusquely, "will you accompany me, please."

Following the major back to an office in the main square of the camp, Pinchuk suspected he was going to be placed under arrest. Perhaps he would be charged with desertion. Or possibly for inciting illegal immigration. Either way he would be court-martialed, and his search for Leah would be over.

But, to Pinchuk's surprise, the major only wanted his help.

"I see they listen to you," he said. "They trust you." Perhaps Pinchuk could convince the Jews that it was safe to go into the camp showers. That they could throw away the rags they were wearing; they would be given new clothes. That they did not have to take bread from the dining room each night and hide it under their pillows. He spoke on, and Pinchuk was moved by the concern in the man's proper British voice.

"I'll try," Pinchuk said when the major had finished. "But after what they've been through . . ." His words trailed off.

". . . I know," said the major.

The two men looked at each other in silence, left dumbfounded by what the human race had done to itself.

It was raining heavily by the time Pinchuk was ready to leave the next morning. According to his map, a camp in Radstadt, Austria, was only forty kilometers away. He decided to go on despite the weather.

But the map, he discovered as he drove, did not indicate that the road through the Katchenberg Pass went up into the Katzberg mountains. The rain soon turned into snow.

Reluctant to turn back, Arie followed the road as it climbed higher, and the snow grew heavier. When the windshield wiper broke, Pinchuk used a piece of wood to clear away the snow. Still he drove on, determined to reach the camp in Radstadt. Even as he continued, he knew what he was doing was irrational. But it also seemed very important. He did not want to stop.

It was not long, however, before he had no choice. The road had disappeared under all the snow. He felt a rush of panic. He was in the mountains, alone in a blizzard. It was quite possible that he would freeze to death. Just then he saw a funnel of smoke rising into the sky. He got out of the jeep and walked toward it.

There were four men sitting around a fire. Parked nearby were two snow plows. They had been clearing the road, but they had stopped for the day. It was pointless, they explained, when it snowed like this.

Pinchuk told them he needed to get to Radstadt today.

They laughed. Impossible, they said. The road out of the mountains would remain closed for days.

Pinchuk pulled the revolver from his holster and ordered them to get on their plows.

Within minutes the convoy of vehicles—two snow plows and the jeep pulling the trailer—was moving uneasily down the mountain. The pace was that of a slow, deliberate walk. But Pinchuk had his gun unholstered, and the plows kept going.

It was one in the morning when Pinchuk saw the lights of Radstadt. The forty-kilometer trip had taken fourteen hours. Nevertheless, he had felt that it was very important to get to Radstadt, and he had succeeded.

A tide of people converged excitedly around Pinchuk when he walked into the camp barracks later that morning. Men shook his hand. Women kissed the yellow star on his uniform. Children tried to touch his gun. Yet above all the commotion he heard a plaintive cry in Yiddish, *"Liebl, das bin ich, Simcha, der katzev."*

Pinchuk turned, and there was Simcha the butcher.

He ran toward him. The two men hugged, sobbing unashamedly, joy and sadness rushing through them.

"Do you know where Leah is?" Pinchuk finally asked, still shaking with emotion.

Simcha said she had been living with him in Reflovka, but then she left.

"Where did she go?"

"Bytom," said the butcher.

"Bytom? In Poland?" Pinchuk asked incredulously.

"Yes, Liebl, your sister's in Poland."

THIRTY-NINE

Carmi returned from Paris wondering if perhaps he had promised too much. He needed the resources of a full-time military unit: trucks, food, blankets, petrol, even border passes. The Brigade could covertly provide the men, but Brigadier Benjamin and his English adjuncts would never sign off on the necessary requisitions. The more Carmi thought about it, the more he realized that it would require an army to move the refugees to the sea. So he invented one. And he did it with just three letters: TTG.

TTG had the short, crisp punch of a military acronym. It sounded like the name of an army unit. But Carmi had chosen the letters from a phrase in a contrived, nonsensical portmanteau language, part Yiddish, part Arabic. The words were *"tilhas tizig gesheften."* Roughly—and it was meant to be rough—translated, it sneered, "Up your ass." But only the Jews from Palestine knew that.

Once Carmi had the name for his new unit, he started quickly issuing orders. Boxes of blank British Army of the Rhine requisition forms, work tickets, and transport authorizations were stolen with surprising ease from the Brigade storerooms. Peltz simply told a clerk, "I want you to pretend I'm not here." And when he walked out of the storeroom with dozens of boxes, he realized he could not reprimand the private for failing to salute. The soldier was, after all, only obeying orders.

At the same time, the call went out for men who had been artists in their former lives, and three reported for duty. Carmi led the volunteers into a room and handed each of them a document with the brigadier's signature. "I want you to be able to sign the old man's name in your sleep," he instructed.

Carmi returned several hours later to see how they were progressing. The three documents were returned to him. He was immediately annoyed. "What is this?" he thundered. "I told you to study them."

One of the men stepped forward. "These are not the old papers," he explained with a smile. "And Benjamin did not sign them."

After that demonstration, the forgers went to work. Dozens of work forms and transport orders were "signed" by Brig. E. F. Benjamin. With the authority of these counterfeit documents, the TTG unit was rapidly outfitted.

British transport units across Europe accepted the paperwork without question and transferred trucks to TTG. Busy quartermasters glanced quickly at the papers and returned with cans of petrol, boxes of fresh food, dozens of uniforms for the men of TTG. Work orders were issued transferring soldiers from the Brigade to special assignment as drivers for TTG. Literally overnight, the farmhouse in Camperosso, once the base of operations for a conspiracy of killers, became the headquarters of a phantom army.

Now that they were mobilized, the TTG units swiftly went off to war. The priority was always the children, and they rushed across Europe to gather them up.

A convoy of two British army trucks heading toward Linz. On their fenders were the unit markings: TTG. Their destination: a DP camp in the American zone.

Peltz and Maxie Kahan were the officers seated in the cabs of each truck. Both men could drawl the King's English and carry themselves with the insouciance of British officers and gentlemen. They were perfect frontmen.

The trucks parked just inside the gate and the two officers walked into the main office. As luck would have it, the U.S. Army camp commander was not around. A sergeant, instead, inspected the papers authorizing the transfer of eighty orphans to the British TTG unit. "They're all yours," he quickly said. And Peltz had the impression that the American could not have cared what else they took as long as all the necessary forms were signed in triplicate.

While Peltz and Kahan completed the paperwork, the trucks drove to the orphans' barracks. Carmi, the TTG driver, walked in and clapped his hands to get their attention. The curious children surrounded him, and he asked in Yiddish, "Who wants to go to Palestine?"

It did not take long to fill the two trucks. But more children wanted to leave, so Carmi explained they would have to "play sardines." "We're going to pack you into the truck like sardines in a can," he warned. But none of the children seemed to mind. "To Palestine, to Palestine," they began yelling once they were seated in the rear of the trucks.

On every operation, two TTG soldiers sat in the back to keep things orderly. This time they could not control the excited noise. Carmi was in the cab and he could hear them. "To Palestine, to Palestine," the children chanted.

Peltz and Kahan climbed aboard, and with a good-bye wave to the accommodating sergeant, they headed back to Italy. Passing through the American, even the Russian zones, was not a problem. Guards glanced at the TTG transport orders and waved the trucks on. The British border guards, however, would be more scrupulous. And there was always the additional danger that the British Field Secret Service agents were on duty. Their primary mission was to stop refugees from boarding a boat to Palestine, and they could be both fierce and relentless in their work.

As the trucks approached the British zone, Peltz pounded twice on the partition between the cab and the rear of the truck. Then he repeated the signal.

The soldiers in the back ordered the children to be quiet. Not a word, not a sound, they sternly insisted. Some of the children started to tremble, and tried not to cry.

The cover story was that TTG was transporting truckloads of prisoners of war to Rome. Peltz had typed transfer sheets, each page bearing the names of dozens of prisoners, complete with rank and date of birth, all affixed with proper seals and signed by Command officers. Usually this was sufficient. Guards never bothered to look under the rear tarpaulin.

But on the way back from Linz, crossing from Austria into Italy, a British MP returned the sheaf of papers to Peltz and challenged, "TTG? I never heard of that unit."

"Really?" Peltz said with unwavering calm. "Well, we're very hush-hush. Strictly need-to-know. Afraid you must not be cleared, old chap."

The MP waved the trucks through.

When the convoy crossed into Italy, Carmi celebrated by singing "Hatikvah." The men joined in, and the children, wanting to be like their new heroes, sang too. All the way to Tarvisio their youthful voices rose up. They were filled with commitment to a land they had never seen, but was all they had.

In Italy, the refugees were brought to the camp in Pontebba. They would wait there for a week, possibly longer, until Arazi completed the complicated arrangements for the next stage of the journey. Eventually they would move farther south to Bologna, then Florence, then to the ships that would smuggle them into Palestine.

But these were busy weeks. As children poured into the camp—four hundred in one convoy alone from Munich—the wait for ships grew longer. And Carmi decided to put these delays to a good use.

Using money from the Haganah in Paris, Carmi leased a decrepit villa from an Italian count and transformed it into a school. Men from the Brigade, a few of them teachers before the war, taught the children Hebrew, Jewish history, the geography of Palestine, and told them about life in the kibbutzim. There were even lessons in farming.

And no less crucial instruction for their new lives in Palestine, they trained the youngsters to be soldiers. Boys and girls were taught how to march and how to shoot. Children who months ago had cowered at the glances of Nazi guards, who saw their mothers and fathers sent to their deaths, learned from Jewish soldiers the skills that would enable them to fight for their futures. The hills around the "Hebrew Hotel," as the local Italians called the villa, echoed each day with the loud reports of gunfire. While at night around campfires, there was folk dancing, singing, and talk about the marvels of the warm rich land they would soon see with their own eyes.

Each child that Carmi loaded into a TTG truck was another life saved. And each had his own story.

Aaron Derman was eighteen and had fought with the partisans during the war.

"I went to Slodim (in Poland) right after I was liberated to see if maybe somebody was still alive. I knew that nobody from my family survived, but I thought maybe somebody else. Like a cousin, or friends.

"When I came to Slodim I couldn't find anybody. It was like a cemetery. It was hell. So I couldn't wait long enough to run out from Slodim and just go.

"But where do you go? We didn't have a family, so Palestine was the family. If I could only find a way to go to Palestine. It was our dream."

He made his way west. He convinced the Red Cross that he was a Greek, not a Jew, and they gave him papers that allowed him to travel from Lublin in Poland to Bratislava, Czechoslovakia. From there he made his way by train to Graz. Then he walked through forests and over the Alps until, exhausted and hungry, he arrived at a British transit camp in Austria. He was detained there with his fifteen-year-old future wife, Lisa.

"We couldn't go anywhere. We were stopped. Held like prisoners. But when we are in Austria these soldiers come for us, real soldiers, and they tell us they are going to take us away. And we sort of, you know, put things together that these must be our liberators. They put us on trucks.

"We drove for quite a while and they told us we have to be very quiet. That there is some danger involved. They had taken us illegally, and we did exactly what they told us.

"At the military control points, you have to lie down on the floor and they covered us with tarps. And nobody is allowed to cough. Nobody to talk.

"But then after we cross the border into Italy they tell us they were from the Brigade.

"Tremendous! We cricd! We screamed! We kissed one another! Can you imagine, from the ghettos and the fires? To see Jewish soldiers?

"It was such a joyous thing. Tears and laughter and crying and happiness. All of it a mixture of everything together. To see Palestinian Jews coming to rescue Jews. It was a very, very emotional time for us. And they showered us with sweets and food. They bring us clothes and whatever they could.

"They were wonderful. They were warm. They were giving. They showered us with everything. With affection. With everything they could. They really treated us in the kind of way that you would expect your brothers to treat you.

"And they told us that we will find a way to Palestine. They told us they will find a road for us. And we believed them."

Yet it was demanding work. During those first weeks when Carmi looked at the faces of the children he saved, he could not help but think about those who had perished. And after he heard a survivor describe how babies were thrown over the heads of their mothers into the gas chambers, he was on the verge of losing faith. If God was in control, Carmi thought, how could He have allowed such evil to succeed?

In time, however, Carmi began to see things differently. He had just returned to Italy from Salzburg with a truckload of refugees, the oldest eighteen, the youngest ten, when it occurred to him that even in the horror there had been miracles. Divine intervention had allowed these children to survive the war. And it was no less a divine act, no less miraculous, that the Brigade had been put in the position—to have the will, the geography, the resources, the cunning—to save them.

But all the while agents of the Field Secret Service were watching the Brigade. And they were becoming very suspicious.

OC FSS Vienna

Secret & Personal

The Unauthorized Movement and Clandestine Activity of Jewish D.P.s in Austria

1. During the past three months, a number of reports have been received concerning the unauthorized movement of Jewish D.P.s into and out of the British Zone of Austria. The nature of these reports has been such as to suggest that this movement is part of a mass emigration from Northern Europe in the general direction of PALESTINE . . .

Investigation carried out to date indicates that the persons who assist the unauthorized movement of Jewish D.P.s [are] . . . personnel of Jewish units . . . they regard themselves as humanitarians and Jews first rather than impartial members of the British Forces.

```
  Action: Discreet measures to prevent units or
individuals in the British Army from facilitat-
ing the contravention of standing regulations by
Jewish D.P.s.
```

In the last week of July, the British took their first "discrete measure." They allowed the refugees to work their way across Hungary and Austria, continue on through the Russian zone at Graz, and then enter back into the British zone.

But once the survivors arrived at the transit camp called Judenburg, near the border between Austria and Italy, the Field Secret Service took control. The FSS guards refused to let them leave. This was as far as the Jews could go. They would not be able to cross into Italy. And they certainly would never get to Palestine.

When Peltz heard the news, he was downhearted. He found Carmi and told him, "That route's closed. It's over. There's no way we'll ever get anyone out of Judenburg again."

FORTY

Carmi listened to Peltz, but he refused to accept that they were defeated. He conceded that he could no longer walk into Judenburg, wave his forged documents, and drive off with a truck-load of refugees. In their new state of alert the British would immediately be suspicious. The risks had, he agreed, escalated.

But he could not allow the survivors to remain trapped. He would not abandon them. For the British to imprison them after all they had suffered struck him as cruel, even vindictive.

Carmi tried to formulate a plan, but his anger only gave him dangerous ideas. He needed to think calmly. As he often did when "things got nerve-racking," Carmi hiked into the hills surrounding Tarvisio. In his solitude, he tried to look at the situation as a soldier would.

He walked for hours. By the time he had wandered back to the barracks, he had his plan. It would be an entirely different sort of operation. And if it worked, everyone in the camp would be long gone before the British realized even a single refugee was missing.

The convoy left Camperosso early the next day. There was a line of twelve trucks, with Carmi driving the lead vehicle. When they approached the transit camp, Carmi pulled off the road heading to the main gate and drove into the surrounding forest. The other trucks followed in single file.

They headed deep into the woods, the trucks bouncing over the uneven ground. Carmi finally came to a halt and the convoy stopped in a line behind him.

Carmi hopped down from his truck. He was dressed in civilian clothes, his pants and jacket mismatched and soiled. The visor of his cap was pulled deliberately low over his forehead. His face bristled with stubble, and this helped to hide his tan.

"How do I look?" Carmi asked Robert Grossman. Grossman had helped extract the list from the Gestapo agent, and he had been part of all that had followed. On this operation, he was Carmi's adjunct.

Grossman studied his friend with a theatrical concentration. Finally he told Carmi there should not be any problem.

"Yes, I should be able to pass for a Jew," Carmi agreed, making a rare attempt at humor. Then he found his commander's voice and spoke to the men.

He would return when it got dark. If he had not appeared by midnight, they were to leave without him. The trucks were to drive out of the forest with their lights off. Once on the road, they were not to stop until they reached Camperosso. No matter what happened, Carmi insisted, no one was to try to rescue him.

Grossman saluted.

Carmi returned the gesture. Then he set off, heading toward the camp, the power in his compact body unmistakable even under the shabby coat.

Carmi had no difficulty entering the camp. The guard did not even raise his head. He was accustomed to Jews strolling in and out of the surrounding forest; there were only so many diversions the refugees could find to pass the long days.

Once he was inside, Carmi went from hut to hut and explained his plan. The refugees listened with excitement. His authority was reassuring, and it allowed the people to believe that what he was suggesting was, regardless of all their other instincts, quite reasonable.

But then one young man asked what would happen to their belongings. And quickly a number of other troubled voices shared this concern.

Carmi had never anticipated this would be a problem. What did they own? A hairbrush? A prayer book? There was nothing that could not be replaced. One girl insisted she could not leave without her goose-feathered pillow. It was absurd.

It was essential to Carmi's plan that the refugees did not reveal they were preparing to leave. If every Jew had a pack strapped to his back, even the most disinterested guard would suddenly be suspicious.

Yet listening to the survivors, Carmi realized he had made a mistake. Their attachment to these small possessions was largely irrational. However, they had lived through experiences far beyond reason. Who was he to judge them? Only the most callow of men could ask these Jews to abandon the few mementoes they had brought out of a world that no longer existed.

So Carmi surrendered. All he could do was hope his decision would not jeopardize the entire operation. He told them to pack, but they were to leave their backpacks and valises on top of their cots. He would make sure they were brought to Italy. "You have my word."

The refugees filled their packs and bags and deposited them on their cots. Then they waited, counting the hours until it would be dark.

"Louder," Carmi instructed. "More noise."

Everyone in the camp, perhaps 150 people, mostly teenagers, was gathered around a blazing bonfire. They were congregated far from the huts, near the treeline that delineated the dense forest. And everyone was singing.

But Carmi was not satisfied. He wanted to make sure the camp commandant, perhaps busy with his evening bridge game or, for all Carmi knew, deep into his gin, heard. He wanted someone to be curious enough to send a few guards over to see what the Jews were up to.

"Come on," he urged, "make some real noise."

They sang louder, clapped their hands, and stomped their feet. And soon the guards approached.

The guards kept their distance, but watched intently. They appeared amused. The Jews were putting on quite a show.

When Carmi casually lifted his hat, the first couple walked hand-in-hand into the woods. When they were gone, another boy and girl headed off. This couple put on a more spirited performance, embracing, their hands all over each other even before they reached the privacy of the trees.

The guards observed it all with wolfish smiles. But after a few more amorous couples wandered off into the darkness, they seemed to lose interest, and returned to camp. There was nothing unusual to report, only boys and girls being boys and girls.

They never realized that while couples went into the woods, none ever came back.

The Brigade waited in the darkness. As soon as a couple entered the forest, a soldier met them. The teenagers, exhilarated by their daring, boisterously congratulated one another. Some had to be sternly warned to keep their voices low, to move quickly. A few couples found it difficult to break off their embraces and had to be separated. But the soldiers were efficient. They led them through the woods to the trucks. And the couples kept appearing.

Around the bonfire, one couple left and the next prepared to go. It was all very orderly, almost mechanical. In time, however, the teenagers grew impatient. They began leaving in hurried clusters.

Carmi started to lose his temper. "This is a military operation," he lectured.

But too many people were already gone. Carmi realized if the guards returned, any explanation would sound improbable. Exasperated, he waved his cap and ordered, "Everyone to the woods."

* * *

Eleven trucks filled with refugees left the forest. The remaining truck waited for Sgt. Fisher and the two soldiers who had gone into the camp to retrieve the backpacks and valises. Carmi had told Fisher that it would not matter that they were in uniform. At this hour the guards would be sleeping. And even if they were awake, they would not pay attention to British soldiers. Nevertheless, just in case, he ordered Fisher to cut the phone lines before they entered the camp. Carmi pointed to a pole across from the commandant's office, and then he was gone.

As Carmi had predicted, there was no sign of the guards. It did not take the men long to gather up the bags.

But as they were about to run into the trees, they were spotted. Stop, stop, they heard voices yell. Wait. Don't leave.

In his charged state, it took Fisher a moment to realize that the words had been spoken in Yiddish. Turning, he saw a group of perhaps forty refugees standing in the dark.

He quickly learned what had happened. After weeks of travel, this band of refugees had walked into Judenburg an hour ago. But they could not find any Jews. Then they saw Fisher sneaking through the compound, and they understood.

They begged Fisher not to leave them here.

Fisher did not know what to do. There was not room in his truck for forty people; if only Carmi had left him one of the big Dodges. Yet even if he did somehow manage to squeeze them in, he did not have any additional travel documents. He would never get them through the border checkpoints.

And there was a more immediate worry. If they went on this way, the noise would certainly wake up the guards. Then no one would be leaving.

He improvised a hasty plan. Tomorrow evening wait in the woods near the camp, he told them. Someone will come for you, he promised. Then he ran toward his truck, hoping the guards had slept through the commotion.

* * *

But someone—a guard? an officer?—must have been awakened by the desperate voices. Near dawn the order was given to inspect the refugee huts.

The guards went from hut to hut, and then rushed to get the commandant.

When the commandant picked up the phone to alert the border guards, the line was dead.

The sun had not yet risen when they reached the border. Carmi, dressed in his TTG uniform, showed a sleepy British MP the convoy's travel permits and handed over the typewritten lists identifying the prisoners of war he was transporting. There were many pages, and each single-spaced sheet listed rows of names, ranks, and dates of birth.

Carmi imagined what the MP was thinking as he flipped through the pages: He could inspect each truck, or he could go back to bed.

"Move on," Carmi heard the MP say. Then he raised the barricade and let the convoy pass.

The convoy reached Pontebba before lunch. As the teenagers climbed down from the trucks, Carmi noticed that he was sweating. His uniform was soaked with perspiration. Odd, he thought. It was not really that hot a day.

By the time he was back in Tarvisio, his skin was burning. He was in bed with a fever of 103 degrees when Fisher told him about the forty Jews still left in the camp.

Carmi tried to rise out of bed, but he could not manage.

"Get me Johanan," he told the sergeant, and he lay back down.

Peltz decided to make the trip alone. The other men were exhausted, and it would be easy enough. If he took one of the Dodges, he could cram everyone into the back. He would drive up to the camp, hurry his passengers aboard, and then he would be off.

When he entered the forest, acting on a soldier's instinct, Peltz changed his plan. He parked the truck quite a distance from the camp and went on foot to the rendezvous point.

He was walking through the woods when a squad of British guards emerged from behind the trees, and pointed their rifles at him. "What is the meaning of this?" Peltz demanded. "I am a British officer."

The men lowered their weapons. He certainly sounded like a British officer. But they had their instructions. They escorted Peltz to the commandant's office.

The commanding officer, a major, was not so easily assuaged. The disappearance of nearly all the Jews from his camp had left him very agitated. "Gangster! Child snatcher!" he yelled at Peltz. "You even dared to cut my telephone line."

"With all respect, sir, I have no idea what you are talking about," Peltz said.

"We'll see about that. You're one of the refugee thieves."

"I think there is a mistake," Peltz insisted. He explained that he had been granted a furlough to look for his family. He had entered the camp hoping to discover someone who might have seen them.

"You're not even a soldier," the major shouted. "Show me your AB64."

Peltz did as ordered, and without concern—his military ID was counterfeit, the name on the card pure invention.

But before the major could examine it, a sergeant entered the room with a flash from Command. He relayed the message word for word, uninhibited by Peltz's presence.

Perhaps he thought it would be of no consequence if Peltz knew. After all, everyone would find out soon enough. Or possibly he thought Peltz would pay no attention; it did not affect him. Or maybe he decided the major would want to know this information right away no matter what else he was doing.

The news immediately refocused the officer's priorities. He would have to deal with the "child snatcher" later. He ordered the sergeant to take Peltz away.

Peltz followed him out of the room. In all the haste, no one noticed that Peltz had taken his AB64 off the major's desk and put it back into his pocket.

The sergeant led Peltz into another office. "Wait here," he ordered, and made sure to lock the door behind him when he left.

Peltz did not wait very long. He went to the window, opened it, and climbed out. The drop to the ground was only a few feet.

He did not run through the camp. He walked slowly and stood erect like the British officer he was, regardless of what name he happened to be using. But when he reached the forest, he ran.

He was back in his truck and on the road within minutes. He pushed the big Dodge as fast as it could go. He was certain Carmi would want to know what he had overheard as soon as possible.

FORTY-ONE

Carmi, still feverish, opened his eyes and saw Peltz standing above him. He had learned long ago that no one ever woke anyone to share good news.

With some effort, Carmi managed to sit up. His mouth was dry and he asked for a glass of water. He sipped it as Peltz reported what he had overheard in the camp commandant's office: The Russians were pulling out of Graz. The British would be taking control of the city.

Carmi, stunned, set down his glass. For the past three months, Jews fleeing Eastern Europe had been directed by the Bricha to the Austrian city of Graz. They came by the hundreds, on crowded trains, and on foot. The Russian colonel who controlled the area was a Jew, and he made sure the haggard refugees were comfortably housed in one of Graz's grander hotels. He also instructed his men not to pay too much attention to the cargo inside British army vehicles leaving the Russian zone on their way west toward Italy. The arrangement had enabled Carmi to smuggle truckloads of Jews into the camp at Pontebba.

But when the British controlled Graz, it would all come to an end. They would cordon off the city. Their stern-faced FSS units would be stationed at every highway checkpoint. Mischievous schemes might rescue a few Jews from a transit camp, but Carmi knew such tricks would be futile against an organized British occupation force.

Peltz waited to hear what Carmi would say. And he noticed that his friend was very pale. Just how sick was he? Peltz worried.

Eventually Carmi spoke. He asked Peltz how many Jews did he think were in Graz, waiting to cross into Italy and go on to Palestine.

Peltz did not know. "Besides," he said, "it's out of our hands. It doesn't matter."

"You must have an idea, Johanan. How many?" Carmi said hotly.

A thousand? Peltz guessed.

At least, Carmi thought. And if he could save one thousand Jews, if he could help bring one thousand pioneers to Palestine, that would be significant. It meant that one thousand lives would have a future.

"When are the British taking control?" he asked.

"Midnight tomorrow."

"Then we have two days," he said.

Carmi got up from his bed and began to get dressed.

Trucks were the initial problem. Carmi had a dozen TTG vehicles parked at the farmhouse in Camperosso, but squeezing a thousand refugees into twelve trucks would be impossible. They needed more vehicles, as many as he could get.

Carmi, although still weak with fever, quickly arranged a meeting with Maj. Shlomo Shamir and several of the Brigade's high-ranking Jewish officers. He told them what was about to happen in Graz, and asked Shamir to authorize the release of thirty additional trucks from units throughout the Brigade.

Carmi was astonished by their response. An operation of this size was too dangerous, the officers argued. When it failed, the Brigade would be embarrassed, possibly even disbanded. And if the Brigade was sent back to Palestine, it was not unreasonably pointed out, there would be no one in Europe to assist the Bricha.

Carmi tried to control his temper. To his mind, it was simply a question of courage and will. There were risks, but they were not insurmountable.

But he could also sense Shamir's growing hesitation. If the chief Haganah official in the entire Brigade ruled that Carmi should not be given the trucks, the operation would be much more difficult. Perhaps even unfeasible.

With great control, Carmi put on a neutral face and said he was withdrawing his request. Then he left the room.

But it was only a strategic exit. Shamir had not formally prohibited him from taking the trucks. He still might find a way.

"Know what I would have told them?" Peltz snapped when Carmi recounted the meeting.

"That's why I'm glad you weren't there, Johanan."

Peltz, however, had an idea. Arie Pinchuk was a captain in the transport company. He could authorize the transfer of the vehicles. Despite the bitterness of their last meeting, Peltz still felt Pinchuk could be counted on.

Except they could not locate Pinchuk, and no one seemed to know where he was. But, luckily, Carmi did find another friend in the motor pool.

Carmi had known Yigal Caspi before the war; they had shared some adventures in the Haganah. Now Caspi was in charge of the 178th, the transport company which had joined up with the Brigade before the Senio offensive.

When Carmi asked him for thirty trucks to use to smuggle Jews into Italy, Caspi had only one question: "Who will be commander of the convoy?"

"I will be responsible," said Carmi.

Caspi immediately agreed to provide the trucks.

After dinner, Carmi gave the order to move out. He was in the lead vehicle, a jeep, with Peltz and Haim Laskov seated next to him. The long convoy—forty-two trucks stretched in a half-mile line—rumbled in response to his command. The big engines came to life. And the race to Graz began.

Their only hope was to leave the city with their cargo of refugees before the British took control. They had until midnight tomorrow—thirty hours from now.

It did not matter that Carmi was sticky with a cold sweat, his throat sore, his every move an exertion requiring reserves of strength. It did not matter that there had been no time to paint over the Stars of David that were on the Brigade's trucks. It did not matter that their travel authorization documents were hasty, obvious forgeries. There were one thousand Jews in Graz, and they had to bring them out before the British locked up the city at midnight tomorrow.

It was a cool, pleasant night and as the convoy rolled ahead Carmi tried to sleep. But he was unable. His fever made him uncomfortable, yet he suspected the reason had more to do with nerves.

The intensity of his investment in this mission was enormous, and it was fueled by a private motivation. He had come to hope that if he could save so many poor souls at one time, he might demonstrate that his own tarnished soul was worthy of being forgiven for the acts he had committed in rage and madness.

He remained focused on these thoughts, on his own hopes, as the forty-two trucks raced through the summer's darkness to Graz.

FORTY-TWO

Around eight in the morning, an hour or two after Carmi finally fell into a deep troubled sleep, the convoy approached the outskirts of the city. Camped on both sides of the road into Graz was an entire British division waiting for the signal to move into the Russian zone.

Laskov nudged Carmi awake.

"What do you want to do?" Peltz asked.

Carmi looked at the rows of British troops assembled in two vast fields and considered his choices.

They could proceed to the first checkpoint and, with luck, enter the city. The sooner they arrived, the more time they would have to gather refugees. But what if the Soviet troops challenged his men at the entrance to Graz and refused to let the convoy pass? Or worse, what if they allowed the trucks inside the city and then made their move. Fighting combat-hardened Russians in the streets of Graz was something he did not look forward to.

Waiting at the roadside and entering the city with a British division would be more prudent, and certainly safer.

Seated in his speeding jeep, Carmi weighed his choices. There were many arguments for caution, but there were one thousand reasons not to delay. He told Peltz to proceed.

The convoy went on toward Graz, while the puzzled British troops assembled in the fields watched in mute wonder.

About ten miles from the center of the city, they came to the first checkpoint. Four wooden horses had been placed end-to-end in a half-hearted effort to block the road.

The jeep halted, and the three men waited to see what would happen. No Russian soldiers approached.

The men exchanged uneasy glances.

Carmi got out of the jeep and began pushing the barriers to the side of the road. Peltz and Laskov covered him, gripping their guns. But the checkpoint was deserted, and the only sound the men heard was Carmi dragging the wooden barriers across the gravel.

When the road was cleared, Carmi signaled the line of trucks behind him to move on.

There were four more checkpoints on the way into Graz, and they were all unmanned. The convoy proceeded without delay.

The Russians have fled, Peltz rejoiced. They've pulled out, he told his friends. Carmi was not so certain. Soviet troops could be waiting in the downtown streets. Until midnight it was a Russian city, and it would be best to take nothing for granted.

A mile outside Graz, Carmi noticed a large park and ordered the jeep to pull over. The convoy followed, and circled around Carmi's vehicle.

Carmi gathered the men and explained his plan. They were to wait here while, accompanied by Peltz and Laskov, he entered Graz, alerted the refugees, and found a location in the city that could accommodate the Brigade's forty-two trucks. When he was done, he would return and lead the convoy to the pickup point. Once the trucks were in place, the refugees could board. With luck, they would drive out of Graz hours before midnight.

He left without mentioning that the success of the plan depended on whether the Russian troops had already withdrawn. But he knew this was not necessary. If they had not, his men would find out soon enough.

* * *

The jeep drove down deserted, silent streets. There were no Russian troops in Graz and, it seemed, no civilians. The quiet was unnerving. Peltz looked around suspiciously. Carmi's hand instinctively reached for his revolver.

When the jeep turned onto the city's main thoroughfare, the noise erupted. Thousands of people lined both sides of the street, cheering, waving Union Jacks, and throwing bouquets of bright flowers. They had come out to welcome the British occupying force.

"Wave," Carmi instructed his shocked friends.

Peltz and Laskov gestured to the crowd. In response, the cheers grew to a roar, and flowers rained down. The jeep drove on, the only vehicle moving through the festive city streets.

By the time the jeep had crossed the city and halted in front of the imposing, fortresslike building that housed the refugees, its hood was covered with a rainbow of petals.

Quickly, hoping not to be noticed, the men entered the hotel. It was deserted. There were supposed to be a thousand Jews living there, but the gloomy building held only dust and silence.

With increasing trepidation, the three went to the second floor. It was empty too. Carmi did not like this at all. Peltz slipped his rifle off his shoulder.

When they climbed to the third floor, Carmi called out in Yiddish. "The Jewish Brigade has come," he yelled through the empty hallways. "We've come to help you get to Palestine."

Within moments people appeared. They hurried down the attic stairs, out of locked rooms and closed closets. They emerged from hiding places throughout the hotel, the walls of the old building echoing as they rushed toward the soldiers. A crowd surrounded the men, throwing excited questions at them. They hugged the soldiers and each other in their joy.

From the midst of this commotion, the redheaded partisan who was the refugees' leader made his way to Carmi. He explained with a grin that everyone had assumed the British soldiers had come to take them to a DP camp. "We thought you were the goys," he said. "But this is wonderful. Wonderful."

The British really will be here soon, Carmi warned. He instructed Red, as the partisan was called, to tell the refugees to pack their bags and get ready to leave.

"How much time will you need?" Carmi asked.

"A half-hour," Red said.

Carmi was more realistic. "Tell them they have three hours," he said. "Then we leave no matter what."

But Carmi still had to find a spot where he could park his half-mile of trucks and load his passengers without attracting attention. He got Red a British uniform from the jeep and told him to put it on. This way, he explained, his presence in a British army vehicle would not attract curious glances. And with the effusive Red as their guide—"This is wonderful, simply wonderful" he repeated to Carmi's growing annoyance—they drove through Graz searching for a secure parking lot.

They found a playing field adjacent to a school. Carmi inspected the site, marching back and forth twice over the grassy plot before finally deciding it was big enough for forty-two trucks.

On the way back to the hotel, Carmi told Red that it was important that the Jews get to the field as quietly, as covertly, as possible. Travel in small groups, Carmi instructed. Use different routes. Secrecy, Carmi reiterated, was essential. Understand? Carmi demanded.

"Of course," said Red when they dropped him off at the hotel. "Nothing to worry about."

Returning to the park where they had left the convoy, the jeep passed the advance party of the British division. It was on the opposite side of the road, coming into the city as they were driving out. Carmi saw the disappointed looks on the faces of the soldiers who had assumed they would be the first British troops to enter Graz. He smiled, and gave them a consoling wave.

* * *

For the second time that day, flowers poured down on Carmi. His convoy had pulled out of the park into a break in the line of British troops coming into the city, only to find they were directly behind the British commander's open touring car. The trucks with the Star of David on their fenders followed the general into the city and the men from the Brigade happily shared his flowers.

But then they broke away. On Carmi's signal the forty-two trucks left the motorcade and headed to the school site.

The men moved rapidly. All the trucks were parked with their engines facing front, their roof tarpaulins fastened, their tailgates down. Carmi was proud of how efficiently it had all been accomplished, and for the first time that day he was beginning to feel confident.

Then he heard singing, voices raised in a Yiddish song, and all his optimism collapsed. He rushed to see who was making the noise. Marching up the street was Red, and he was leading a formation of at least a thousand people, packs strapped to their backs.

Carmi stared in disbelief, and he felt blood rise to his head. "Quiet!" he shouted. "Enough!"

He ran down the block and grabbed Red. The startled man began to cower. But as Carmi tugged at the lapels of Red's coat, as his anger was about to goad him out of control, he became aware of the refugees crowding around. He could see their confusion and their distress. At once Carmi felt embarrassed by his rage. He removed his hands from Red's coat, fixed him with a hard stare, and returned to his men.

Later, as the soldiers loaded people into the trucks, he was even able to joke about it. Imagine how pleased the British general must have been, he told Peltz, to see a parade of grateful Jews taking to the streets to welcome him.

There were five checkpoints on the way out of the city. It was nearly eleven that night when the convoy pulled up to the last one. British soldiers were gathered under the lights, waiting impatiently for the stroke of midnight. Still, Carmi was unconcerned. The Royal MPs were powerless for another hour. The convoy would make it safely out of the city.

Then a horn sounded. Carmi glanced back and saw a truck on the side of the road. He ordered his jeep to turn around.

At once, the trucks behind him came to a halt. The long line of vehicles stopped about a quarter of a mile from the last checkpoint out of Graz.

When Carmi's jeep pulled over to the side of the road, an agitated soldier told him there was a problem. Carmi followed him to the back of the truck. He shined his torch inside and saw a woman lying on the floorboards in a puddle of blood.

"We don't do something," the driver said, "she's going to die."

Carmi checked his watch: 11:15.

"In the police, in Haifa, I learned first aid," said Laskov. "I could try to help her."

Carmi saw the trucks in the middle of the road and yelled to Peltz, who was still sitting at the wheel of the jeep. "Get everyone across the border. Hurry."

Peltz turned the jeep and raced back toward the head of the convoy.

Carmi told Laskov to do what he could. Then he looked up the road. The convoy had once again started to move. The first truck was pulling up to the checkpoint. The British soldiers did not stop it. *That's it,* Carmi thought. *Don't hesitate. Go!* The first truck drove through the checkpoint and out of Graz. And the long line began to follow.

Carmi glanced at his watch: 11:30.

He went to see about the woman. "How is she, Haim?"

"Bad," Laskov said.

"Can she travel?"

Laskov shrugged.

They were running out of time, Carmi knew. Midnight was approaching. He would have to make a decision. Suddenly he ran to the cab of the truck, climbed in, and ordered the driver to get back onto the road. "Hurry," he shouted.

With Carmi in the front seat, the vehicle sped through the checkpoint. It was the last truck to leave Graz. Carmi looked again at his watch: 11:52. They had made it with eight minutes to spare.

The convoy arrived in Pontebba shortly before breakfast. A medic told Laskov he had saved the woman's life. Carmi was about to ask for aspirin for himself, but then he decided he did not need it. His fever had broken, and anyway he felt fine. Better, in fact, than he had in months.

FORTY-THREE

Pinchuk, meanwhile, prepared for the next stage of his journey. He had started out four weeks earlier believing that with luck he would find Leah in Italy. If that failed, he had been confident he would eventually find her in a camp in Germany or Austria. But Simcha the butcher had said Leah was in Bytom, Poland, and now Pinchuk was resigned to making the long trip east.

That night in the camp in Radstadt, Pinchuk spread out his map and tried to plot the most direct route. And as he studied the flimsy rectangular sheet, its profusion of colors and dotted lines that differentiated not only a jumble of countries but also a maze of military zones, he began to grow daunted.

His destination was deep inside the Russian zone and he would be a British soldier traveling without a visa. The Communists might decide he was a deserter, or a black marketeer, or even a spy. Jail, he imagined, would be one punishment. A Red Army firing squad would be another, and no less probable.

Before falling asleep that night, Pinchuk decided it would be too reckless to enter Poland without an entry permit. In the morning, he would head across Austria to the Soviet delegation in Salzburg. Yet even as he traced his new itinerary on the map, he knew there was no logical reason to believe the Soviets would want to help.

* * *

Salzburg had loomed in Pinchuk's mind as grand and refined, the dowager birthplace of Mozart. But driving through the crowded streets, he experienced only a nasty, scurrying postwar city. Whenever he stopped for directions, someone tried to sell him something. Watches, tires, even women were offered. And Pinchuk's refusals did not end the conversations. Was the gentleman a seller, then? Did he have cigarettes, nylons?

Pinchuk was simply looking for the headquarters of the American Jewish Joint Distribution Committee—or Commitat as it was known—that monitured the DP camps throughout Europe. The organization, he had been told in Radstadt, had taken over an abandoned building and might be willing to give him a room. He made one wrong turn after another as he tried to follow the directions offered by the street corner blackmarketeers, until finally he located the block-long building with its red tile roof.

The couple—Pinchuk suspected they only lived like husband and wife—that ran the Commitat were delighted to welcome a soldier from the Jewish Brigade. They quizzed him exhaustively about Palestine, the White Paper's restriction on land ownership, the situation with the Arabs. In return they offered him a firm bed and, a true pleasure after all his vagabond weeks on the road, a hot bath. And later they fed him lavishly.

It was a dinner rich with aromas and tastes that were immediately familiar, the essences of an earlier life. There was borscht, the tart soup thick and beety; pierogi, the large doughy dumplings stuffed with a coarsely ground meat; and, when he was sure he could not eat another bite, they set a steaming plate of earthy brown veal stew in front of him. Each taste was a memory, a remembrance of what he had lost. And in its evocative way, an articulation of what he was setting out to reclaim.

For the first time in weeks, he slept contentedly.

The next morning, freshly bathed and shaved, smart in his captain's battle dress, Pinchuk arrived at the Soviet delegation. Seated at the desk in the entry hall was a female Soviet officer, perhaps twenty-five years old, blond and blue-eyed. Her uniform fit her very well. Despite her attractiveness, Pinchuk sensed she was a cool, detached, unsympathetic sort.

When she addressed him in Russian, Pinchuk simply shrugged. His instincts told him it would be to his advantage if he did not reveal he spoke Russian. "Good morning," he responded with a baffled grin.

To his surprise, she switched to a thickly accented but fluent English. "What do you need from us?"

Pinchuk explained that he was a British officer looking for his sister. He needed the Soviet delegation's assistance to get to Poland.

"We do not handle such matters," the woman said. "We are here on another mission. I cannot help you."

Arie tried to persuade her, but the longer he continued talking, the more he feared his attempt at presenting a calm, well-reasoned case was unraveling. He was nearly shouting, but he could not help it. In his pounding heart the stakes were clear: If he did not find some official way to travel east, all would be lost.

"What is the matter?" a voice interrupted.

A Russian officer was coming down the stairs. He wore his uniform trousers and was absently fastening up the buttons of his white rubashka.

The woman gave a dismissive yet accurate summary in Russian explaining who Pinchuk was and his reason for wanting to go east.

The man looked at Pinchuk.

Pinchuk did not speak. In truth, he did not dare. But Pinchuk fixed his eyes imploringly on the officer. He wanted the officer to understand how important this was to him.

"Listen," Pinchuk heard the man at last say in Russian to the woman. "Do you remember that Col. Polovnik is supposed to come by this afternoon on his way from Innsbruck to Vienna? Maybe he'll take him."

The woman began to argue, but this only served to resolve the issue in the officer's mind. He ordered her to tell Pinchuk to come back at three that afternoon to meet the colonel. When she had finished explaining everything in English, the officer placed a reassuring hand on Pinchuk's shoulder. *"Dagravilis,"* he said. *We're set.*

"Thank you," Pinchuk said. "I am so very grateful." And his undisguised joy made any further translation unnecessary.

Pinchuk returned just before three. The female Russian officer directed him to a salon crammed with upholstered furniture. She pointed to a grass-colored sofa against the wall and told him to sit.

The officer deliberately found a chair on the opposite side of the room and, a few minutes later, was joined by another female officer. Dressed in civilian clothes, she was dark-haired and a bit fuller than her friend; and she, too, carried her weight well. The two women spoke about men and bedroom matters in Russian as if alone.

Pinchuk was surprised by the coarseness of their words and the directness of their observations. It was soldiers' talk, but Arie had never suspected that women spoke or thought that way.

Occasionally the dark-haired woman would point to Pinchuk and ask her friend if she was sure he did not understand Russian. "No, no," she would say.

Pinchuk sat on his side of the room, his legs crossed, smoking Players, trying to keep an absent look on his face. He felt as though he were a spy.

After an hour had passed he began to worry that the colonel had not appeared, but he also found it difficult to concentrate on anything other than the conversation he was overhearing. His need to find Leah had been so intense, so blinding, that he had nearly forgotten there was a world of experiences and emotions beyond his own concern. He wondered if this was what a normal future might offer if he could complete his mission.

In time a door opened and the officer he had met that morning and Col. Polovnik, a genial, disarmingly handsome man, entered. Polovnik shook Pinchuk's hand, and after introducing himself in a slightly hesitant German, said, "To Vienna."

"To Vienna," Pinchuk agreed.

* * *

Pinchuk sat in the back of the Opel, and the colonel sat with his driver in the front. The two spoke in Russian. They were talking about gasoline shortages when Pinchuk dozed off.

It was a long trip but, except for three flat tires, uneventful. When the colonel spoke to Pinchuk, it was in his stumbling German.

At around eleven that night they approached the Danube. All of Pinchuk's weariness seemed to vanish when he saw the lights of Vienna in the distance. He felt exhilarated by the prospect of completing one more leg in his journey.

Then the colonel told Pinchuk to hide on the floor of the car. There were guards on the bridge and Pinchuk's name was not in his travel papers. Pointing to his chenille coat, he told Pinchuk to get beneath it.

"British army officers do not lie down on the floor of cars and hide," Pinchuk said, immediately regretting the words and his imperious tone. It was something Peltz might have said. But he felt it would be a mistake to conceal himself. It was certainly demeaning and, if he were discovered by a guard, possibly dangerous.

The colonel became very angry. With the indignant air of a man whose kindness had gone unappreciated, he warned Pinchuk that they all would be in jeopardy if Pinchuk was arrested. But Pinchuk would not cooperate.

As they drove onto the bridge, the Russian ordered his driver to stop. For an uneasy moment Pinchuk thought he was going to be told to get out. But instead the colonel exited the car, and walked the twenty-five meters to the sentry hut.

From the backseat, Pinchuk watched the colonel hand the guard a cigarette and a sheaf of papers. The Russian kept his body between the car and the guard, blocking the soldier's view. Moments later when the guard returned the papers along with a crisp salute, Pinchuk finally relaxed.

The colonel climbed into the Opel, and he quietly ordered the driver to proceed. Then he turned to Pinchuk and announced that he would have to deal with the American guard at the other end of the bridge on his own.

Pinchuk was taken aback. He had not realized there would be Russian *and* American guards on the bridge to Vienna. Perhaps, he thought, it would have been wiser to hide on the floor under the chenille coat. But he answered in German, "All right. I'll manage."

When they reached the American checkpoint, the Russian handed over his papers. He also made it clear to the GI that the man seated in the back was not included in his permit.

The guard motioned for Pinchuk to roll down his window. "Are you Russian?" he asked in English.

Pinchuk explained in his own slightly accented English that he was an officer in the British army serving in the Jewish Brigade.

"You are not registered," the GI said.

Pinchuk could not refute that charge since it was true. Instead, he told him everything. He spoke about his search for his sister, about his arduous weeks going from camp to camp, about how he desperately needed to get to Poland. He wanted the American to understand what he was trying to accomplish, as well as what it meant to him.

"I can let you pass," the GI decided. "But I can't guarantee they'll let you back out."

"Please," said Pinchuk. "Just let me pass."

It was after one in the morning when the Opel finally entered Vienna. A short distance from the bridge, the colonel ordered his driver to stop, then turned to Pinchuk. "Get out," he said harshly. "At once."

Pinchuk exited, and the car swiftly drove away. He stood there wondering what to do. After a moment, he began to walk down the dark, empty street.

FORTY-FOUR

TOP SECRET
MESSAGE OUT
From: G.H.Q.
To: Troopers London

GUARD (.) TOPSEC (.) Evidence exists of illegal immigration to PALESTINE through ITALY and indication of organization for movement of displaced persons wishing to immigrate illegally into PALESTINE involving Palestinian troops. Consider in order to make definite and effective break in chain of communications it is essential quickly to remove from this theatre all Palestinian units. . . .

Two days after Carmi's convoys returned from Graz, the Brigade was ordered to leave Tarvisio. The British Command had begun to realize some of what the Jewish soldiers had done to assist the refugees, and they decided that if the men were stationed far from the Mediterranean, they would no longer be a concern.

On July 29, 1945, the Brigade, six hundred trucks and five thousand men, began a long trip west to their new camps in the Low

Countries. Like Hannibal and his invading troops, they went over the Brenner Pass, down to the fairytale valley of Innsbruck, and across Austria. Then they drove into the gray gloom of Stuttgart, and entered the interior of Germany. Over the next two weeks they traveled through a bleak procession of bombed-out cities, past the bricks and rubble that had once been the foundations of the Thousand Year Reich.

It was a dangerous, stressful journey. Five thousand Jewish soldiers were in enemy territory. Many of the men once again chalked defiant slogans on the tarpaulins of their trucks: *"Deutschland Kaput! Kein Reich, Kein Volk, Kein Führer! Die Juden Kommen!"* Trucks veered out of line to smash into men bicycling on the other side of the road. At night, soldiers looking for revenge sneaked into villages, assaulting women, burning houses. For others, the hatred was beyond words or deeds. "The deeper into Germany the more silent we all became," Gerald Smith, a British-born member of the Brigade, explained. "There was utter devastation everywhere. People stared at our Jewish flags with incredulity. We pitied the people of ruined Cologne until we encountered that indescribable smell and realized that we were close to the Bergen-Belsen concentration camp."

For Peltz, the voyage through Germany was an opportunity to see his own harsh journey in a purer light. He had come to Europe to claim his inheritance, the prodigal returning to all he had left behind. But there was nothing. Zabiec was in ruins. And his parents, his grandfather—their graves were not even marked.

Nevertheless, Peltz was no longer imprisoned by his sorrows. He had succeeded in working his way beyond grief. In the course of his service with the Brigade, on the battlefield and in the covert missions, he had fashioned a new identity. In Palestine he had been a Pole, an outsider and a private man. In Europe he had realized he was a Jew, and his destiny was intertwined with the fates of his brothers and sisters. He had found something substantial to take the place of what he had lost: a people, and a land.

For Carmi, the trip through Germany was also one of relentless emotion. That he completed it without succumbing to vengeful acts left him with a restored pride. It was proof, he wanted to believe, that he was once again worthy, a knight ready to go off to fight the battles he was certain history would soon require.

By the end of August the Brigade had settled into their new camps. Brigadier Benjamin established headquarters near Brussels, and units spread across Belgium and Holland. The British had moved the Jewish battalions far from the Mediterranean, but this was irrelevant. It was "family business," as Carmi had called it. And there was still much to be done.

In Belgium, Carmi found his new billeting had its advantages. After the unit's English sergeant major was sent home, Carmi was promoted to regimental sergeant major of the Second Battalion. The promotion was gratifying, but more important was the authority that went with it. Carmi now had control over the battalion's vehicles and manpower roster. He immediately brought the TTG units up to strength. And he assigned men—David Littman, Netanel Lorch, and others—to establish a school for the children who were still forced to live in the horrors of Bergen-Belsen.

There were just a few makeshift classrooms in dank, claustrophobic Barracks 41, but for many of the children living in the camp it was the beginning of their return to normalcy.

Ze'hava Brumberg had been a teenage prisoner in a series of camps, and each day, tired and exhausted from work, thirsty and hungry, she escaped into a fantasy. In her imagination she focused on a world of teachers, pupils, books, and notebooks. Beyond the barbed wire fences she was certain there were girls who walked to school every day, laughing and holding schoolbags. More than anything, she had a naive, unrealistic wish to go back to that wonderful period of her life when she was at school. She wanted to return to a time when she could concentrate on solving math problems, and not worry about the day's allotment of thin soup running out before she was served.

When the war was over and the soldiers from the Jewish Brigade came to Bergen-Belsen, it was as if her long-held dream had finally, incredibly, come true. "They brought," she rejoiced, "a young spirit and ideals. The school brought us together into one family. They strengthened our self-confidence and showed us we are no less than other youths who had not spent time in the concentration camp. School became the home that revived our souls."

But Carmi was above all a military man, and it did not take him long to discover the soldier's treasure that surrounded him. Dozens of British Army of the Rhine armories, warehouses crammed with arms and ammunition, were located near the Brigade's new camps.

Meanwhile, the news from home grew more disquieting. The Attlee government would not compromise its policies on immigration and statehood, and the Yishuv, after the searing lesson of the Holocaust, also felt it had no room to make concessions. Events were moving forward with a grim momentum. Lord Moyne, the British minister of state, had been assassinated in Cairo by Jewish gunmen. Riots had broken out in Tel Aviv. Six Jews were killed by British bullets. The colonial government threatened to impose a twenty-four hour curfew on the entire Jewish community. If that happened, the Yishuv promised the British would have to place 600,000 Jews under arrest. Bevin could make his cracks about the Jews pushing to the head of the queue, but when the battle over Palestine erupted it would not be a shoving match. The soldiers from the Brigade would once again be on the front line. The only question was whom they would be firing at—the British, the Arabs, or, no less likely, both.

Without guilt or hesitation, Carmi ordered the men to take whatever they could from the British armories. Oly Givon repeatedly strolled into one warehouse and each time he left carrying sniper rifles still in their cases. Peltz could not believe the crates of heavy machine guns and bazookas that seemed, at least in his mind, to have been left just for his men's taking. Shaul Ramati, an officer, was more circumspect. The armory under his command always held the correct number of Enfield rifle cases. The crates, however, were filled with towels. "What can I tell you?" the Polish-born and Oxford educated Ramati said. "There was a higher morality I was obeying."

It was a treasure trove, and it was no challenge to mine. But they still had to get the weapons to Palestine.

In the fall of 1945, a sergeant major from the British army requisitioned a large windowless warehouse in a lonely corner of Antwerp. MPs were stationed by the door, and the guard was kept around the clock. A regimental signboard identified the inhabitants to any curious passersby: TTG.

Inside, Carmi supervised dozens of busy soldiers. Throughout the day trucks arrived and men unloaded their booty of stolen weapons. The armaments were carefully packed into barrels and welded shut. As a final touch, the barrels were painted with a bright Red Cross.

As the TTG's armory grew, Carmi sent a flurry of messages to the Haganah Command in Paris. Get us a ship, he wrote at least twice each week. We'll make sure the cargo gets on board.

At the end of the third week, Munya Mardor, the pale blond Haganah officer in charge of arms acquisitions, arrived in Antwerp disguised as a British sergeant. Arazi had bought a 400-ton wreck called the *Tel Hai*, he informed Carmi. In ten days it would be ready to sail to Palestine.

Carmi asked where it was docked.

Marseilles, Mardor said sheepishly.

Carmi had assumed the boat would be anchored in an Italian port and his convoy would make the long wintry trip across Europe over familiar roads. He was unprepared for a journey to France.

But he threw himself into this new challenge. Carmi pored over the maps with the men who would be driving the trucks, while a separate team searched the French-Belgian border for the least conspicuous place to cross. Soon the logistics fell into place. At the end of the week he gave the order to start loading the three hundred Red Cross barrels onto the trucks.

But just as the heavy work got under way, Carmi received an urgent summons to the apartment on rue de Ponthieu in Paris.

There was a problem, Shadmi announced when Carmi arrived. The British were pressuring the French to expel the *Tel Hai* from its waters. The ship would have to leave for Palestine right away.

Carmi said there was no need to worry. His men would get the arms on board in time.

Shadmi cut him off. Carmi did not understand. The *Tel Hai* could make only *one* trip from France.

Carmi was still confused.

There was a refugee camp outside Antwerp, Shadmi went on. It housed about 1,500 inhabitants. Mostly young men and women. Survivors, without families.

Now it was clear: it was a choice between sending guns or children to Palestine. Carmi agreed that Arazi would have to find him another boat.

But Shadmi was not done. There still remained the question of how to get the refugees from Antwerp to Marseilles. They did not have visas and they had to be moved across Europe as soon as possible or the opportunity would be lost.

"My men will do it," said Carmi.

"I was counting on that," said Shadmi.

Midnight. A drizzle dripping through the thin fog. One by one the trucks pulled up to the camp on the outskirts of Antwerp.

The passengers were waiting, jostling one another, their packs on their backs. As soon as a truck arrived, soldiers helped the children on board. Boys sat on one side, girls on the other. Each vehicle held about thirty people, and when a truck was filled it pulled out and joined the line. After an hour, the convoy was ready.

When Carmi had begun planning the operation, he did the mathematics and determined he would need sixty trucks. Even if he managed to "borrow" so many vehicles, that large a convoy—almost a mile of trucks—would proceed at a snail's pace. The journey to Marseilles would stretch on for three, possibly four days. For the survivors, crammed into the back of a truck, still weak, it would be torture. In the end he decided to take only thirty trucks. And his men would make two separate trips, thirty hours each, one right after the other.

"Forward!" Carmi yelled just after one A.M. on March 9, 1946, and the convoy moved out. Once again, he was in the lead jeep. Peltz rode in a truck this time; it was decided his fluent English would be valuable if the vehicle was stopped.

Carmi had measured out a good part of his life in tense border crossings and he suffered a familiar edginess as they approached France. But no sooner had his jeep pulled to a halt than the sentry waved it on. A gift of five cases of whiskey from a French-speaking Brigade captain had convinced the guards that the convoy was transporting German POWs to a work detail.

Once they passed through the border, Carmi almost relaxed. Every two hours the convoy stopped at the field kitchens that had been set up in advance along the carefully mapped route. The grateful refugees were given a warm cup of tea and a chance to stretch their legs, and then they were hurried into the trucks for the next leg of the journey.

As they moved down from the chilly hill country, the weather worsened. All night a lazy fog had floated in the air, but just before dawn it grew thick and dense. Carmi had a difficult time seeing the road in front of him. And then he heard the sound of metal smashing into metal.

Three trucks had rear-ended one another. The cabs of two of the trucks were smashed, and fluids oozed onto the dark roadway. No one was hurt, but the vehicles were immobilized.

"Here's what we do," Peltz announced in his definitive way after he had spent only a few moments bending over the damaged engines. "Get ropes and we'll tow them."

Carmi was not sure this would work, but the alternative was to leave ninety refugees stranded by the side of the road.

"It's worth a try," he told Peltz.

* * *

In that limping way, they drove on. It was night when they reached the windy port, and entered the camp Mardor had set up a few days earlier.

"Welcome to Marseilles," he greeted Carmi.

An hour later, Carmi, fortified by a thermos of black coffee, was on his way back to Antwerp. And thirty hours after that, he was leading the second convoy toward southern France.

"Welcome to Marseilles," he heard Mardor repeat two days later.

Carmi was so exhausted he could not even get out of his jeep. To his embarrassment, he needed to be lifted from the vehicle and carried to the bed Mardor had found for him.

Carmi tried to remove his boots, but he fell fast asleep before he could undo the laces. He was still sleeping soundly the next evening when the *Tel Hai* and its 1,400 passengers moved out to sea.

FORTY-FIVE

The Russian colonel had left Pinchuk in an unpopulated area of Vienna, the buildings shuttered and lifeless. Lost and exhausted, Pinchuk wandered in the early hours of the morning through the bleak city streets.

Suddenly he heard a rumbling noise and looked up into the bright yellow headlights of an oncoming car.

Pinchuk stepped out of the light and unfastened the flap on his holster. If it was the military police, they would want to see his travel permit. If it was crooks, they would want his money. His heart pounded as the car stopped and the window was rolled down.

A familiar voice asked in German: "Do you have a place to sleep?" It was the colonel's driver. Still cautious, Pinchuk approached the car. There was no one else inside.

"No," said Pinchuk.

"Didn't think so." The Russian started laughing, and that confused Pinchuk. "Get in," he said.

When Pinchuk was in the warm car, the Russian explained he had been impressed by the way Pinchuk had stood up to the colonel. " 'A British officer does not hide on the floor of a car under a coat,' " he quoted. "Good for you. Bravo." So after he had dropped the colonel at a hotel, he went back to assist the proud British officer.

The man did not explain where they were going, and Pinchuk did not ask. Even a cot in a crowded barracks would be an improvement over his earlier prospects.

They drove to a large gray stone apartment building with wrought iron balconies in the central part of the city, the Kertner Ring. The Russian parked in front, and used a key to open a tall set of doors decorated with elaborate ironwork. He led the way through the lobby and up a curving staircase. At the landing he fumbled with another key, pushed open an apartment door, and entered. Pinchuk, baffled and curious, followed him in.

It was an extraordinary apartment; one vast room seemed to lead into another equally grand. But Pinchuk was most impressed by the lavishness of the decoration. There were mirrored walls and a clutter of furniture that to his eye seemed delicate, expensive, and decidedly feminine. As Pinchuk looked around, he reminded himself of the erratic course his evening had taken. One moment he was walking the streets, the next he was in a gaudy palace. The lesson was that he could not allow himself to despair.

He asked the Russian whose apartment it was.

"The colonel's," the man said with a laugh. His trysting place. The colonel, however, had business with other officers at the hotel, so he had given his driver the key.

Tonight, the driver went on with undisguised glee, we get to sleep in feather beds and drink the colonel's vodka.

The two men had almost finished off the first bottle when Pinchuk decided the proper moment had come to ask for the driver's help in getting to Poland.

"You need to get a permit." The Russian poured them another drink.

The two men raised their glasses toward one another and drank up.

The Russian emptied the last of the vodka into their glasses. "You need to go to Command. In the Hotel Imperial."

Once more they drained their glasses in a gulp.

Pinchuk looked at the man with level eyes. He wanted to convey the seriousness of his need.

"All right," the Russian said, shaking his head in mock exasperation, "I will take you."

Then he went off to find another bottle. And from somewhere in the sprawling apartment, Pinchuk heard the driver cackle with amusement, " 'A British officer does not hide on the floor of a car under a coat.' "

In the morning Pinchuk's head was pounding, but he rose early and, in preparation for his meeting, shaved and polished his combat boots. He could only imagine how the Russian was suffering. But true to his word, the man drove him to the Hotel Imperial, shook Pinchuk's hand with great ceremony, wished him luck, and then quickly left.

Guarding the entrance were two huge Russian soldiers cradling Star rifles with their distinctive round ammunition magazines, and Pinchuk approached them. They refused to let him enter. Travel permits, one of them barked, were issued behind the hotel.

Pinchuk followed the grudging directions and walked through an alleyway into a courtyard crowded with people. A line of bodies snaked to a short, husky Russian officer. There was a table and chair for the officer, but apparently he preferred to stand.

Pinchuk took his place at the end of the long line and found it moved with an ominous speed. The Russian neither spent much time on anyone nor, if the brusque exchanges Pinchuk had overheard were any indication, issued permits.

"Go home," the officer admonished the old woman in front of Pinchuk. "We do not lose things around here."

Pinchuk had prepared a small speech. He had hoped to reach out to the compassionate Russian soul, a national trait that had seemed real enough in the pages of Tolstoy. But after what he had witnessed this morning, he no longer expected sympathy.

Nevertheless, Pinchuk approached the officer without hesitation. He held himself erect, a British captain at attention. The Russian stared at his uniform, and Pinchuk hoped it was not too wrinkled. He was about to make his appeal when the Russian spoke. "Are you a Jew?" he asked.

"Yes," Pinchuk snapped, ready for a confrontation.

The Russian offered his hand. "So am I. What do you want?"

Pinchuk started to tell him about his search for Leah, but the Russian interrupted him. Details were unnecessary. He spoke to the stunned Pinchuk as if he were discussing a trivial matter. "Wait. I will finish with these people. Then I will take you to lunch at the officers' hotel and I will arrange everything for you."

An hour or so later the two men were having borscht in the Grand Hotel. General Rokovsky sat at the same table, and at first was eager to question Pinchuk about Palestine. But after Pinchuk said the British would not be able to maintain their control, the general seemed disturbed and withdrew from the conversation.

When the long lunch was finally over, the officer introduced Pinchuk to a bearded blond major in charge of issuing travel permits.

This time Pinchuk told his story in detail and when he was done, the major considered it carefully.

Please, Pinchuk prayed silently, a permit is just a piece of paper. That's all I'm asking for.

The major finally spoke. In two days there would be a transport train leaving from Braunau carrying Polish soldiers east to Katowice for repatriation. "I can get you on that train."

The Polish officer in charge of the train told Pinchuk to hand over his revolver. It would be returned when he got off.

Pinchuk looked around the compartment. The seats were crowded with Polish soldiers and he had a Star of David on his coat. He felt their resentment, and he hoped they felt the anger that seethed in him.

"No," he told the man. "Ridiculous." He moved his hand to his holster flap as if he were preparing to draw.

The Pole held his stare, but finally walked away.

The trip took two days. It was night when the train stopped at Bytom. Full of anticipation, Pinchuk went directly to the camp where Simcha had seen his sister. But Leah was not there.

Pinchuk was not overly discouraged. He had managed to get to Poland. He was confident he would find his sister. It was, he reminded himself, simply a matter of not losing hope.

For the next two weeks, he searched DP camps and nearby towns. In the camps there was a board for messages. The notes were only a few lines on scraps of paper, yet they told stories filled with an immense sadness: "Moshe Abramovitz looking for his wife Sara, his son Mendel, his daughter Rachel." There was nothing from Leah.

People wanted to help. "Yes, she was here," one man reported, "but she left three months ago." An old woman from a town near Reflovka said Leah had died of typhoid and was buried somewhere in the woods. No, another woman told him, she was certain Leah had gone to Moscow.

Each frustrating day it became harder and harder to maintain the belief that he would ever find his sister. With every new disappointment, more of his self-confidence seemed to seep from him.

After three weeks, he decided he could not bear to continue. Poland was suffocating; it was as if the dust of all the cremated Jews still floated in the dank air. He had to escape. He would return to Salzburg. Defeated, he would rejoin the Brigade.

While Arie was traveling east, Leah was making her way west. She had been in Bytom, but she, too, could not bear to stay in Poland. She met up with a group of young refugees and traveled with them in a series of crowded trains to Czechoslovakia, then Budapest, and finally to Austria, arriving in Vienna days before her brother. They spent three days at the Rothschild Jewish Hospital.

As they were leaving, a representative from the Commitat in Salzburg urged them to write their names and next destinations on a card. If a relative was searching for someone, perhaps the Commitat would be able to help. Leah knew no one was looking for her; her only relative was a world away in Palestine. But her friends were writing their names on the small white cards, so she did, too. Where was the harm? she asked herself.

FORTY-SIX

In the first week of April, nearly a month after his return from Marseilles, Carmi was summoned once again to Paris. The three hundred Red Cross barrels—tons of weapons—had already been transported to an interim warehouse in Toulouse, and he had been waiting to hear if a boat to Palestine had been arranged.

Carmi arrived at the apartment on rue de Ponthieu and was led into a cell-like bedroom adjacent to Shadmi's office. A tea kettle whistled, but no one paid any attention. Shadmi came into the room, and gestured to a narrow bed with a maroon cover. Sit, he instructed. He would not be much longer. Then he left.

For a while, Carmi sat there ignored. He looked out a filmy window into a sunless courtyard and listened to the voices in the next room. The words were indistinct, but the tone was agitated. He sensed a crisis.

At last he was escorted into the office and directed to a straight-backed chair opposite Shadmi's desk. Tea was served, and Carmi sipped the hot, sweet drink. He was certain he had not been called to Paris to arrange a routine arms shipment.

The news Shadmi reported was very disturbing. Arazi had found a boat in the Italian port of La Spézia, but had been arrested with one thousand illegal immigrants on the way to the SS *Fede*.

The Italian police were moving Arazi to a jail in Milan. For the time being, they had accepted the alias he was using. But once Arazi was identified, Shadmi said, they would turn him over to the British.

Carmi knew what would happen after that. Arazi was wanted for stealing arms from the British army. When the FSS had caught up with another accused arms thief, Eliezer Goldberg, they had beaten him methodically and brutally. Carmi doubted the FSS would treat his friend with any less viciousness. Then they would hang him.

And Carmi did not have to be told how Arazi's arrest would affect the covert emigration of Jewish survivors from Italy to Palestine. The entire operation would come to a halt, perhaps permanently.

He sat rigidly, suddenly very alert, considering all the implications. The effect on the Yishuv of the loss of an essential man. The effect on a family of the loss of a husband and father. A picture filled Carmi's mind: his dapper, cavalier friend tied to a chair, bloody and beaten.

The old man got up from behind his desk and approached Carmi. He stood erect, and Carmi instinctively rose to attention too.

"What I am about to tell you has already been approved by Moshe Sharret and the leaders of the Yishuv," Shadmi announced gravely. Then he ordered, "Get him out, Israelik. Do whatever you have to do, but get him out."

It was a difficult mission to plan. Would it be launched against the Italian police or the British army? Was Arazi in a city jail or a military prison? All Carmi knew for certain was that he was going back to war.

The next day in Belgium, Carmi assembled his raiding party. On this mission, he decided, speed and stealth were more crucial than size. He selected only a few men, veterans of the *huliyot*, soldiers who had proven their ruthlessness, and their commitment. There was Peltz, Grossman, Sammy Levy, and a handful of others. Reviewing his team, Carmi realized they would need someone who spoke Italian, and he added Mardor to the list. He signed on eagerly.

They left that night. A thousand miles separated the jail in Milan from their base in Belgium and they needed to cover the distance swiftly. They wanted to strike before the Italians identified the criminal locked in their cell. If Arazi was handed over to the British it would be another sort of mission. A raid into a military stockade would be a firefight.

It was a small convoy. There was a pickup truck loaded with the fuel they would need. Then Carmi's jeep. And attached to the jeep was a trailer. The trailer held their arsenal: tommy guns, boxes of grenades, a heavy machine gun, and two mortars.

They passed through Brussels and into France. From the open jeep, the men looked out at staked vineyards, and smelled the rich, pungent earth. They did not stop for meals. "The fewer people we meet," Carmi had told them, "the fewer witnesses there'll be."

Near Epernay, Carmi noticed that the fuel truck was no longer following them. But he was reluctant to stop; it would, he believed, eventually catch up. He asked Peltz to tell him when it reappeared.

Peltz watched the blackness behind them. Time passed, but he saw no sign of approaching headlights.

Finally the jeep turned back. They had not gone too far before they saw the glow in the night sky. Now they drove faster, and at last they rounded a bend and found the pickup lying on its side in a field, enveloped in flames. Behind it, in the unnatural illumination, they could make out dark, swerving tire tracks cutting across a curve in the road.

A crowd had gathered. Grossman, who knew a little French, learned that the passengers had been removed. He kept asking if they were alive, but either his French was inadequate or, the soldiers somberly realized, none of the onlookers was willing to break such sad news. At last one of the Frenchmen pointed to an inn across the way. The three soldiers hurried over.

The bar was surprisingly crowded. The fire had been good for business, Carmi thought angrily.

"Here!" Peltz yelled from a small room adjacent to the bar. Carmi rushed in and saw the driver of the truck. He sat in a chair and blood dripped from his scalp like paint from a brush. But he was alive.

"The others," Carmi asked. "Where are they?"

The driver motioned vaguely toward the stairs on the opposite side of the room. They led to the cellar, and Carmi had a single thought: this was where the bodies would be kept until they were buried.

Yet as all his hopes sank, Carmi saw Mardor coming unsteadily up the staircase. His clothes were covered in blood. His eyes had a faraway look. Carmi called out to him, but Mardor did not answer. Each step requiring great concentration, he walked past Carmi and into the adjacent bar.

Carmi followed and saw Mardor appraising the pretty woman serving drinks. In fluent French, Mardor asked for a cognac, then smiled charmingly at the barmaid before he took a long swallow. He turned to Carmi. "I think the best thing is for me to get a doctor right away." A moment later he collapsed against the bar, and Carmi rushed toward him.

The three men in the cellar as well as the driver should not be moved, the doctor said. He was French, yet he knew enough German so that Carmi and he could communicate. The men were badly injured, but with rest they would recover. Mardor's condition was more precarious. He had extensive internal injuries. He should be hospitalized immediately.

Carmi listened, and thanked the doctor for his help. The doctor said he would return to look at the men the next day. Carmi paid him in advance with pounds sterling, and the doctor seemed quite pleased.

It was only when the doctor was standing at the door with his overcoat buttoned that Carmi took his arm and gently guided the man into a corner. He wanted to speak without anyone overhearing his words.

"It would be most unfortunate," Carmi said, "if the military authorities or the police heard about the accident."

The doctor looked at Carmi thoughtfully. "It is understood, Monsieur."

* * *

The next day when the doctor arrived, he found his patients resting on cots in the cellar. But he could not find Mardor. The doctor's inquiries were answered only with shrugs and, perhaps remembering the warning from the menacing sergeant major, he let the matter drop.

At that moment Mardor was propped up uncomfortably in the backseat of a jeep. He was driving with the remaining war party—Carmi, Peltz, and Grossman—to Italy.

They had lost half their men, and all their reserve fuel. But they had their trailer loaded with weapons. And Arazi was still in jail. No one had considered abandoning the mission.

Earlier, Carmi had called the apartment in Paris. A member of the Haganah command was on his way to the inn to make sure the injured men would be well-treated. When they recovered sufficiently to travel, he would arrange their return to their units in Belgium.

Mardor, however, had insisted he was feeling better, and Carmi chose to believe him. He was the only one who spoke Italian.

The men left after breakfast. To make up the lost time, Carmi planned a more direct route. He decided to go through Grenoble.

It was a mistake. The mountain roads leading into Italy were snowed under. They tried one route, but had to turn back. Another road brought them closer to the border, only to wind up leading to an impenetrable wall of frozen snow. Finally, they found a lower pass that had been cleared and were able to make their way through the Alps.

But the cold and the unsettling drive, the jeep constantly bouncing over icy roads, had been too much for Mardor. He was deteriorating. When he tried to speak, his words were broken up by the rattling of his teeth.

The men exchanged anxious glances.

Carmi tried to keep Mardor talking, to keep his mind engaged. He wanted to keep Mardor connected to life.

Mardor fought on, too. He tried to stay in the conversation that

Carmi was patiently leading him through. But soon he slipped into some place more remote, and lost consciousness.

A shot of whiskey revived Mardor. He was lucid and he managed a brave, rakish smile. But by the time they found a mountain inn, he was fading again.

They nursed him through the night. He was wrapped in a cocoon of blankets, and still his body shivered.

Carmi thought it likely that Mardor would die that night. In his mind, he saw his friend being lowered into a snowy grave in the French Alps. The prospect of an internment in this frigid land, a resting place so unlike the warm earth of Palestine, added to the tragedy. He could not bring himself to leave Mardor's bedside.

Sometime after dawn, Mardor asked for another whiskey. Carmi poured a measure into an enamel cup and Mardor sipped it slowly. With effort, he managed to finish what was in the cup. He was weak, but he was alive.

They carried him to the jeep and propped him up in the backseat. He leaned on Carmi and tried to smile.

They arrived at the Haganah's safehouse in Milan that night. Mardor had already been put to bed when there was a loud rapping at the door. The men looked at one another. No one was expected.

Peltz, gun drawn, covered one side of the door, Grossman the other.

They nodded at Carmi, and he opened it.

Yehuda Arazi walked in.

"Prisons," Arazi said, smiling mischievously, "were made for men like me to walk out of as we please."

FORTY-SEVEN

Both Carmi and Arazi were completely surprised. Neither of the men had suspected who was on the other side of the apartment door. But for the moment the two friends did not ask questions. They simply hugged each other forcefully.

Later over a bottle of red wine, Arazi gave an account of his escape. As he told it, the circumstances were more comic than cunning. One moment he was surrounded by a squad of police. The next they had become distracted, and he strolled off. Carmi suspected Arazi was downplaying the danger, but he and his men laughed along. After the sustained tension of the past two days, they were glad to celebrate.

When Arazi learned that the Haganah had dispatched a team to rescue him, that his comrades had rushed a thousand miles at great risk, that Mardor, his injuries at last properly tended, lay convalescing in the next room, he became very emotional. Tears welled in his eyes, and then he broke down and cried openly.

As the evening went on, however, the talk turned more serious. While Arazi had managed to walk away from the Italian police, the 1,014 refugees were not as fortunate. They were prisoners. For the time being they were held aboard the *Fede*, but soon the authorities would transport them back to the DP camps. The opportunity to smuggle a thousand Jewish survivors into Palestine had been lost.

But even as Arazi described the hopelessness of the situation, Carmi suspected his friend was toying with them. He doubted Arazi had given up, and he was right.

Very tentatively, Arazi began to hint at a new plan. He asked, What if we could turn failure into victory? Suppose there was a way to take advantage of our predicament? What if we could, he said, "wage a public campaign" for the people on the *Fede*?

Carmi was doubtful. The world had stood by while European Jewry had been annihilated. Why would they care now about a thousand ragged refugees?

The British Press Information Office, Arazi argued, could not censor the news in Italy as they had done in Palestine. The story of a homeless, victimized people trying to reach the only community that would take them in would make headlines. It would be, he predicted, an international drama.

Arazi prowled the room while he spoke; it was as if the power of the idea taking shape in his mind left him electrified with excitement. The British, he said, would not allow their proud Empire to be revealed as the cruelest of bullies. Fixed in the condemning stare of world opinion, His Majesty's government would be forced to retreat. Arazi stood still long enough to pound on a table and make a promise: "One thousand refugees will be allowed to leave for Palestine. And this time it will be *legal* immigration."

Carmi was unconvinced. From what he knew of the world, governments, like people, responded only to force. Peltz agreed. It was a bayonet charge that drove the Nazis from La Giorgetta, not a newspaper article. He excused himself and went to bed.

But Carmi stayed up with Arazi. As they talked, Arazi's strategy continued to form in his mind. He suddenly had a new idea and he shared it with his friend: tomorrow morning he would return to the *Fede*.

Carmi protested that it would be too dangerous, but Arazi cut him off. There was something else: he wanted Carmi to come with him.

"I can talk," he told Carmi. "But if we have to fight, it would be good to have you on board."

Carmi was dressed in a pair of baggy pants and a jacket that barely stretched over his chest as he walked with Arazi toward the dock in La Spézia. A few hours earlier he had said his good-byes to the others. After arranging for Mardor's continued medical treatment, Peltz and Grossman would make the long trip back to Belgium and the TTG units. There were still convoys of refugees to transport south and arms to be acquired.

Carmi looked down the sloping street and saw the squads of police guarding the fenced entry to the boat. Once again Arazi was risking arrest by the British, and Carmi was filled with admiration for his friend's courage. But at the same time, Carmi was also curious to see how the resourceful Arazi would persuade the Italian police to let them board the embargoed boat.

Arazi chose a sympathetic face from the row of men in uniform, and explained that he and Carmi were Jewish refugees who wanted to join their people on the *Fede*. The policeman ordered them to leave.

Carmi tried to catch Arazi's eye. If he punched the guard, Arazi might be able to sneak aboard during the confusion. Arazi, deliberately ignoring him, repeated his request to the policeman. This time his tone was more plaintive and imploring.

Again, the policeman refused. But Arazi was undeterred. He begged and pleaded. Finally, the harassed policeman told Arazi that they could do whatever they wanted as long as he would stop bothering him.

As the two men hurried up the gangplank, Arazi shot Carmi a triumphant wink. But Carmi did not offer any congratulations. From his perspective, all they had succeeded in doing was to get themselves locked in a crowded, dangerous prison.

* * *

That afternoon an Italian gunboat tied up to the *Fede*.

The lookouts had seen the warship approaching, and that gave Arazi time to prepare. When the sullen sailors armed with machine guns took their positions on the crowded deck, Arazi strode up to their commander.

People who had proven they could survive the guns of the SS and the Gestapo, he announced, would not be intimidated by these weapons. Remove the sailors, he insisted. It is grotesque, shameful even, to keep unarmed women and children at gunpoint.

The Italian commander apologized; he was only following orders and could do nothing.

"I insist the guards be removed," Arazi repeated.

Carmi waited belowdecks. Earlier Arazi had made him boat commander, and Carmi had wandered among the refugees to select his security force. He recruited three dozen men, mostly partisans who knew how to fight, as well as a handful of surly but steely-eyed teenagers. They were armed with two rifles, one revolver, several knives, and a supply of wooden planks. "At the sound of gunfire," Carmi told his force, "we charge."

On deck, Arazi continued to challenge the Italians. He wanted the guards off the boat.

And in time the commander surrendered. Regardless of his orders, he said he could not allow sailors under his authority to hold women and children at gunpoint. With an air of embarrassment, he led his men off the *Fede*.

That night, still exhilarated by his first victory, Arazi passed a scribbled note to a sympathetic young Italian journalist who had spent the day waiting by the gangplank.

"We are 1,014 Jews," he had written, "refugees remaining after the German massacre, and we are headed for Eretz Yisrael, our natural homeland. We are returning to Eretz Yisrael from which we were expelled 2,000 years ago. There is no power that can prevent us from getting to our destination even if many of us pay with our lives. . . .

"Put an end to our suffering!

"The ship's doctor tells us there are 150 pregnant women on board. Keeping them here under difficult conditions may lead to disaster. We hold those in charge of our delay here responsible for the consequences."

Arazi wanted his message sent as a telegram to the Italian prime minister, the American president, the pope, and the editors of dozens of international newspapers.

It was an impressive list, but Carmi doubted the note would have any effect at all.

Carmi was wrong. The day after the telegram appeared in the newspapers, the British decided to take control of the situation. Maj. Hill of the Intelligence Service boarded the ship with a squad of soldiers and offered an ultimatum: Leave the *Fede* immediately, or my men will drag you people off.

Arazi stepped forward from the crowd on deck and introduced himself as "Alon." He explained to the major that he was the group's spokesman and that he, too, had a statement to make: The moment a British soldier laid a hand on a refugee, the ship would be blown up. He would kill everyone on board, soldiers and Jews, before he allowed his people to be returned to the camps.

Arazi pointed across the deck. Standing by a pyramid of fuel barrels was a broad-shouldered refugee with a match in his hand.

"It will take only one match," Arazi said.

Carmi raised the match high so the British officer could see it.

The sea of people on deck suddenly turned quiet. The British soldiers glanced nervously at the major, awaiting their orders.

The major looked at Carmi. Then back at Arazi.

He ordered his men off the boat.

But that night a British destroyer anchored alongside the *Fede*. And tanks sealed off the quay.

In the morning, Arazi addressed the refugees: "We cannot fail. If we are defeated here the whole illegal immigration movement will be finished."

He announced that a hunger strike would begin immediately. It would continue until the *Fede* was allowed to sail to Palestine.

A banner hung across the gates leading to the dock. In Hebrew, English, and Italian, it proclaimed, "A HUNGER STRIKE OF 1,014 PEOPLE."

At its center was a circle where the number of hours the strike had lasted was posted.

Curious La Spézia residents were the first to congregate by the gates. Italian reporters quickly joined them. Then the French. By the second day, reporters representing English and American papers were also in La Spézia filing stories.

Soon another circle appeared on the banner. It showed the number of immigrants who had fainted from hunger. This number, too, increased each hour.

On the third day Arazi sent a message to the press:

"We are in high spirits despite the 48 hours of our hunger strike. Today 30 people passed out but those who fainted continued to strike.

"Don't wait till there are dead people on the boat. Demand its immediate departure and send protest telegrams to the authorities in Rome."

Carmi was desperately hungry. He tried, however, to keep his mind focused on his responsibilities as boat commander. He had none of Arazi's faith in the efficacy of a "public campaign." Eventually the British troops would storm the boat. He needed to be sure his security forces would inflict as many casualties as possible before they surrendered, or were killed.

He spent a great deal of time instructing the teenagers. They wanted to be soldiers and he could think of nothing more important than teaching them all he knew. If he did not return to Palestine, the thought that they would take his place was a comfort.

* * *

"Sixty-three hours to the hunger strike. Pregnant women are getting weak. Tens of people are fainting. Are you waiting for us to die? If you have any humanity, stop this mass suicide."

Arazi's communiqué appeared on the front pages of newspapers throughout the world.

> H.R. Wishnograd, writer
>
> La Spezia 24 (Overseas News Service)
>
> I just made an illegal visit to the 1,014 Jewish refugees . . . .
>
> The interior of the boat is divided into two parts, separated by a corridor barely wide enough for one man to pass. There are rows of 7-tiered bunk beds lined up next to each other . . .
>
> This ship is somewhat similar to a Chinese river boat and looks unfit for a 2-week sea voyage. The fact that the refugees are willing to put themselves in danger in such a trip shows how desperate and courageous are these people who went through hell before they reached this threshold of the promised land. And even it was taken from them . . .
>
> This morning after the Zionist flag was flown from the main pole, the passengers held a meeting. The decision was unanimous. Continue the strike to the end . . .

Toward the end of the third day, seventy hours after the hunger strike had begun, Arazi sent a new message to the press: Unless their demands were met, ten refugees would commit suicide on the deck every day. The first ten had already volunteered. The "operation" would begin at noon tomorrow.

Five hours later, two black British Embassy sedans arrived at the dock. Maj. Hill, a representative of the British ambassador, and

Harold Laski, chairman of the British Labour Party, had come to negotiate.

"Would you, Mr. Laski, or any of you gentlemen," Arazi challenged as soon as the meeting began, "agree to live in a camp?" Then without waiting for an answer, he turned to Laski and asked, "When did your family reach England?"

Laski seemed annoyed by the question. It was both intrusive and impertinent; he was an Englishman, not a Jew. His ancestors had emigrated well over a hundred years ago.

"If they had not," said Arazi, "you would have been one of us. If the Nazis had not murdered you first, of course."

Now that the tone of the meeting had been established, it did not take long to reach an agreement: Laski would return to England and negotiate with Bevin and Attlee on the fate of the *Fede*. During this interlude neither the British troops nor the Italian police would board the boat. And with that the hunger strike ended.

While they waited for a response from the British ministers, the first night of Passover was celebrated. Last year Carmi had conducted a seder in a deep trench by the Senio River as German artillery thundered above his head. Tonight he sat in the cool night air on a boat with a thousand refugees, a British destroyer anchored nearby. Perhaps next year he would be back in Giv'at Hashlosha alongside Tonka and Shlomit, with Arabs lurking in the wadi beyond the fence.

The episodes in his life were part of a continuum, a struggle that began on this night over three thousand years ago when his ancestors had left Egypt with the hopes of returning to the Promised Land. And as the seder prayers were chanted on the deck of the *Fede*, Carmi felt an immense satisfaction that it had become his journey, too.

On May 19, 1946, thirty-three days after the refugees had boarded the boat, the *Fede*, renamed the *Dov Hos* after a Jewish labor leader, slowly steamed out of La Spézia harbor. All 1,014 Jews would be legally immigrating to Palestine.

A band played on the dock as hundreds of Italians gathered to wish the passengers a safe journey. Carmi, on land for the first time in weeks, walked through the loud, festive crowd aware of his debt to Arazi. He was thankful not only for what his friend had accomplished, but also for what Arazi had taught him. There had been a time in Carmi's life when he had thought violence was the only reliable way to release the rage locked within him. Even when that had passed, he still looked at life with a soldier's logic: one had to storm the gates. But Arazi had shown him that there were other effective methods, and what had worked in the port of La Spézia was also part of the soldier's way.

He hurried back to Belgium and the Brigade with these thoughts in his mind.

FORTY-EIGHT

Pinchuk returned to Salzburg three weeks after he had left. He arrived at the Commitat building after everyone had gone to bed. In the morning he rose early and went to retrieve his jeep from the military garage where it had been stored. He was glad not to see anyone. When he had first come to this city he had been so confident, but now he felt simply a profound disappointment.

He had ventured everything—his life as a soldier, his future as a man—on the hope that he could find Leah. Only he had failed, and now there was no way he could ever come to forgive himself. His life would always be in chaos.

His jeep fueled, he returned to the Commitat for his pack. As Pinchuk was heading down the stairs from his room, the man who ran the organization stopped him.

"Captain, thank God," he said. "I was hoping I would see you."

He led Pinchuk into his office, and handed him a small white card. Written on it in childish block letters was a name: "Leah Pinchuk." And beneath it: "Binder Michal."

Binder Michal Ziedlung was a DP camp near Linz, the Commitat head explained.

Pinchuk did not respond. He continued to study the card,

scrutinizing each word, looking for the proof that his sister's hand had written those letters.

Arie drove wildly. He had searched for Leah for sixty-three days, and lived with the hope for much longer. His anticipation was almost unbearable. Better to pretend this was just another improbable lead, another in his long series of misadventures. But while he tried, he could not maintain this detachment.

Pinchuk arrived at the camp shortly after two in the morning. Dozens of squat gray buildings spread across flat acres of land. Seeing no one, not even a light, he drew his revolver and shot it in the air. A few windows opened, and the camp inhabitants called down angrily to the inconsiderate soldier. The windows were slammed shut before Pinchuk could get in a word.

Arie slept in the jeep, rising with the sun. He drew some water from his radiator and used it to shave; he did not want to look disreputable. When he met his sister, he wanted Leah to be proud of him.

After he was done, he searched the camp. Each building contained dozens of apartments that had once housed SS officers. A list of the current residents was posted near the front door and he used his army flashlight as he scanned the pages.

He went to five buildings before he found his sister's name. Next to it was a room number.

Leah was brushing her teeth when she heard a knock on the door. She assumed it was the girl from across the hall. "Come in," she said.

Arie entered wordlessly. A young blond-haired woman stood with her back to him.

"Leah . . ."

The woman did not move, and she did not speak.

Finally, she turned and saw a British officer.

"Liebl?" she said.

"I have come for you," he said more formally than he had intended.

"Oh my God! Oh my God!"

She ran sobbing into her brother's arms.

They cried. They talked. And they cried some more. She told him what had happened to their parents, but she was reluctant to describe her escape, and how she had managed to survive. He told her about the telegram that had been forwarded from Moscow and how he had gone off on his quest. He did not tell her that he had given up hope. And he did not reveal that the successful conclusion of his long, consuming mission left him overwhelmed with gratitude for the two treasures he now possessed: the gift of her life, as well as the gift of his own salvation.

When they were ready to leave, Leah told her brother it would only take her a few moments to pack. She had another dress, a blouse, and, most valuable of all, a matchbox full of sugar cubes.

Pinchuk told her to leave everything. She did not need it.

Leah hesitated, but then she understood. Her brother would take care of her from now on.

She held on tight to his arm as they walked out of the gray building.

FORTY-NINE

When Carmi returned to his battalion at the end of May, he found himself thinking more and more about home. He was tired. He had fought a war on the front lines only to move without pause into a series of tense, clandestine adventures. He longed for quiet days and nights of uninterrupted sleep.

His thoughts, too, were influenced by his experience on the *Fede.* Much of his time had been spent training the teenagers to fight, and this led him to consider his relationship with his own daughter.

Shlomit had started school, Tonka had written in her last letter. She was six years old, and in those six years—her entire life—he had been away for months, then years at a time on missions for the Yishuv. He could not discuss his covert work with his daughter. And even if he were able, was that any consolation for the many birthdays and Sabbaths they had spent apart? The war in Europe had been over for a year. Carmi wanted to be a father again. And a husband. He was needed at home.

Peltz, too, was eager to return to Palestine. A new ambition now energized him. He wanted to create another Zabiec. Under the searing sun of Palestine, anchored in its rich ancient earth, he would build his homestead.

It would be neither as vast nor as grand as its inspiration. He was no longer a rich man's grandson. But, unlike Zabiec, it would be permanent, a gift that would pass down for generations in an unbroken chain. In this new land, no one would ever be able to steal it away. He waited impatiently in Holland for the day when the Brigade would be deployed, and he could begin his new mission.

But he had seen things in Europe that were too difficult to forgive, or ever to forget. And when circumstances offered him one more opportunity for vengeance, he took it.

Peltz's company was guarding a German parachute regiment in Bloemendaal, Holland. For a man who had been to Mauthausen, it was goading work. One spring morning Peltz selected the fittest soldiers, officers as well as enlisted men from the prisoners' compound. Without a word of explanation, the Jewish soldiers led them at gunpoint into three trucks. A half-hour later, the apprehensive prisoners were unloaded in front of the old synagogue of Haarlem. The men from the Brigade handed out rags, mops, and brooms.

"The floors, the windows, the brass," Peltz bellowed at the Germans. "I want everything spotless."

Pinchuk's trip back to the Brigade's camp in Holland with his sister had been, in its emotional way, another adventure. They sat side by side in his jeep and, making up for the lost years, they talked almost without stop. This time Leah told Arie everything—about hiding in the hole with their mother, about the moment she had decided to run for the woods, about Boris, the swamp, and the partisans. She told him about being alone, terribly afraid, and still refusing to surrender. He was filled with anguish at what his sister had endured, and with awe that she had persevered. And he understood that he had no more important duty than to justify the faith Leah had placed in him.

When he reported back to the Brigade, he began making arrangements. He asked Carmi to secure Leah a place on the next illegal ship going to Palestine. Carmi learned there was a converted cattle ship leaving from Marseilles, and a contingent of Brigade soldiers escorted Leah from their camp all the way to the dock.

Arie did not make the trip to France. He was at sea on a British troop ship. As soon as he learned there was a boat for Leah, he had requested a furlough home. The compassionate Gofton-Salmond quickly approved it. Pinchuk's plan was to find a place for his sister to live, make certain his enrollment was secure at the Hebrew University Law School, and report back to the Brigade.

It did not work out that way.

FIFTY

It had been politics which had persuaded His Majesty's government to send a brigade of Jewish soldiers to fight the Nazis, and it was a political decision that sent them back to Palestine. Their departure was one more consequence of what the Attlee government bitterly referred to as "the immigration question."

The voyage of the *Fede* had been a turning point for both sides. Its success encouraged the Bricha to launch more boats crowded with refugees. In response, the Royal Navy intensified its blockade of the Palestinian coast. And now the captured illegal immigrants were escorted at gunpoint into a Palestinian detention camp at Athlit.

To the Yishuv, the images of squalid refugees being menaced by British might—leaky tubs harassed by a flotilla of destroyers, women and children imprisoned behind barbed wire fences guarded by swaggering troops—evoked the still raw memories of Jews being brutalized by the Nazis. Such disproportionate force was itself a provocation, and the infuriated Palestinian Jews struck back. The Haganah raided the Athlit camp and freed 208 prisoners. They attacked the Lydda railway station and bombed the Haifa refineries. Police launches were sunk and coastal radar stations were destroyed.

The British response was unyielding. They sent more armed troops to the territory. Armored cars loaded with policemen raided the kibbutzim in search of weapons. British paratroops prowled the streets of Tel Aviv, stopping people at will and demanding to see government identity cards. And Palestine moved closer to a state of revolt.

Chief Secretary's Office
Jerusalem, Palestine
TOP SECRET

. . . the ENZIO SERENI, a new wooden motor vessel of about 500 tons, was intercepted by the Royal Navy . . . with 911 illegal immigrants on board and brought into Haifa.

Their retention at Athlit will doubtless be stigmatized by the Jews as "intolerable provocation" and there will be consequent incidents! In fact last night the Givat Olga coast-guard station was again blown up with resulting injuries to 15 British soldiers and one British and one Arab policeman and an unsuccessful attempt was made to blow up the R.A.F. radar station at Haifa. Both these incidents were presumably intended as demonstrations in regard to the immigration question.

As the tensions mounted in the Middle East, the British began to suspect that moving the Brigade from Tarvisio to the Low Countries had not changed anything. The smuggling of Jewish survivors into Palestine would not be stopped until the Brigade was removed from Europe.

These battalions were British troops and they had fought as comrades in the war, but that war was over. And, His Majesty's government realized, a new war had begun.

In the first week of June, 1946, the Brigade was ordered home.

* * *

Two days after the announcement was made by the War Office, Carmi traveled once again to the apartment in Paris. He was filled with happiness at the prospect of returning to his wife and his daughter, and he had no doubt the other men were anticipating their own reunions with excitement. But he also knew that the Brigade's activities in Europe—transporting refugees to the boats, acquiring weapons, teaching children in the DP camps—had to continue.

Shadmi listened to Carmi's idea and, after laughing at its audacity, finally agreed it might possibly work.

And so a group of soldiers from the Brigade went into the DP camps and began their search. They looked for survivors with a certain shade of hair, a certain height, a certain color of the eyes. They were hunting for boys who resembled them.

They found 138 of these "Doubles," as they became known, and smuggled them out of the camps to a country house near Ghent. Here they acquired the biographies of the soldiers who had chosen them. They learned the names of their new fathers, mothers, girlfriends. And they were trained as British soldiers. They learned whom to salute, how to dress, the proper responses to an officer's questions.

The rigorous training lasted two weeks. When it was over, they changed places with the soldiers who had selected them. The Doubles were given Brigade uniforms and identity papers. They would be the ones returning on the troop ships to Palestine.

The 138 soldiers from the Brigade were given new lives, too. The Brigade's forgers had quickly manufactured identity cards with distinctly English names. Men replaced the mezuzahs around their necks with crosses. Photographs of their girlfriends were inscribed with dedications to their new aliases. In their disguises, the 138 men moved out across Europe to continue their secret missions.

It was not feasible for Carmi or Peltz to remain behind. Carmi was the regimental sergeant major and Peltz had been promoted to major. They were well known to the British; not even the most gifted Double could have managed those impersonations. And besides, both men were eager to get home.

* * *

Pinchuk never returned from his furlough. He was ordered to remain in Palestine while his deployment papers were processed.

And soon Peltz's name was called. On the ship to Alexandria, the public address system continually played the Andrews Sisters singing "Drinking Rum and Coca-Cola." Unlike the voyage to Italy nearly two years earlier, on this crossing there was no fear of torpedoes; life jackets were not issued. Nevertheless, it was a tense voyage.

A few of the soldiers were infected with lice, and an order was given that each man have an examination before the ship docked. Peltz joined the line of half-naked soldiers and patiently waited his turn. As he approached the doctor he noticed that the arm of the man in front of him was tattooed with a concentration camp number. He whispered to the man, "The doctor will see." But the Double seemed not to understand the implications of Peltz's concern. "Auschwitz," he whispered back with a philosophical shrug.

Peltz turned to the two soldiers standing behind him. "I want you men to start a fight," he ordered. They looked at him with some puzzlement, but finally said, "Yes, sir." And after the first punches were thrown, as men moved in to pull the two combatants apart, Peltz led the confused Double away.

The ship was also carrying five hundred British paratroopers on their way to Palestine. They were young tough soldiers, and they moved in resentful groups about the boat. Peltz could feel their arrogance. They saluted Peltz when he passed, but he knew he would not be in their army for much longer, or ever again on their side.

Carmi's homecoming was uneventful. The day he was discharged in Rehovot he reported to the Haganah command. He told them about the men he had left behind, and the shipments of weapons they would soon be expecting. When Carmi finished, Israel Galili, the head of the National Headquarters, asked him to choose his next assignment. Carmi was not interested. He had been working for the Yishuv for ten years, since his days with Wingate and the Special Night Squads, and now he wanted to rest. The time had come to live as he had always hoped to live. He was finally going to become a farmer.

He returned to Giv'at Hashlosha. He listened with astonishment and gratitude as his daughter read aloud to him for the first time. He slept next to his wife, and felt the comfort of her presence. He worked long days in the fields in the hot sun. The promise that had brought a boy from Danzig to this new land was at last fulfilled. His years of service had been rewarded.

On a warm day that fall, just three months after his return, Carmi was on his tractor plowing the Nezla fields when he saw a man walking toward him. Despite the distance, he recognized the man's snowy head of hair.

"Leave me alone!" Carmi shouted.

Galili continued toward him.

"Go away!" Carmi yelled.

Galili came closer.

Carmi stared at the approaching figure. And as Galili walked across the field, Carmi understood that all his feelings over the past three months had been simply wishful. Fate had charged him—his generation—with the responsibility of bringing an ancient covenant to fruition. Until this was accomplished there could be neither rest nor rewards. For men like him there was only duty. And the sustaining faith that his sacrifices would ensure that his daughter could grow up as a free woman in a free land.

Carmi turned the key on his tractor and silenced the machine. He jumped down and walked over the freshly turned earth toward his friend, ready to receive his new mission.

PART VI

ISRAEL, 1948 / ITALY, 1995

FIFTY-ONE

On the fifth of Iyar, 5708—May 14, 1948—the independent State of Israel was proclaimed. After two thousand years the Jewish homeland in Palestine was restored.

And even as Israel's new prime minister, David Ben-Gurion, made the announcement, Egyptian bombers took off for Tel Aviv and the armies of the Arab world marched across the borders of the new state.

"There are fifty million Arabs," Ibn Sa'ūd, king of Saudi Arabia, proclaimed. "What does it matter if we lose ten million people to kill all the Jews? The price is worth it."

The War of Independence had begun. Once again the men from the Brigade were called up to the front line. Once again they went into battle waving a blue-and-white flag with the Star of David. Only now it was the flag of their country.

Pinchuk had been teaching English at the Ahad Haam Boys' School and studying law at night, but when war broke out he joined up immediately. In the Brigade he had learned how to fire a mortar, and that knowledge was invaluable to the new Israeli Defense Force. He became deputy commander of a mortar battalion. It was an active unit, and he fought in battles all over the country.

The fighting in the Negev was particularly fierce, and there were many casualties in the drive to push the Egyptians out of Kis Faluja. A corporal in the unit was badly wounded and was sent back to Tel Aviv to be treated. A month passed before Pinchuk found an opportunity to visit him in the hospital on Balfour Street.

He walked into the ward and saw that his friend was being attended by a pretty blond nurse. "This is the life," Yehuda told him bravely. "Have you ever seen a nurse like this?"

"Yes," said Pinchuk as he kissed his sister.

Peltz did not need more than a quick look at the map to understand the importance of his mission. With the Arabs controlling the airport road, Tel Aviv, the entire country actually, was cut off from the outside world. His unit was ordered to take the hill overlooking the road and drive the Arabs out. "Punish them," Peltz was instructed.

There was no opportunity. Peltz led his men up the hill and the Arabs ran. While his men set up an old Austrian artillery piece, a Schwarzlose, he tried to call the company commander to inform him that they had taken their objective. The wireless transmission would not go through.

It was crucial that Command learn the airport road had been secured. Peltz decided the radio might be more effective if he transmitted from a higher elevation. He looked around and saw a flowering sycamore tree, and started to climb.

He pulled himself to a perch in the highest branches. He inhaled the fresh sweet smell of the yellow flowers wrapped about the limbs of the old tree and looked out at a vista of smooth, undulating land. For several moments he sat motionless, lost in thought. When he finally tried to contact the base, he succeeded on his first attempt. Then he climbed down.

He inspected the machine gun installations and, satisfied, returned to the tree. He studied the sycamore with some concentration, and then found a metal bottle opener in his pocket. He nailed it to the tree with the butt of his revolver.

"What are you doing that for?" his puzzled second in command asked.

"When the war's over, I want to find this tree," Peltz explained. "I'm going to build my house right here."

The operation code-named Ten Plagues was designed to break through the stalemate in the Negev. If everything went according to plan, by the time the final plague was inflicted, the Egyptians would be pushed back across the border.

Carmi and his commandos were the first plague. His jeep unit was armed with mounted machine guns and their assignment was to take an Egyptian position by surprise, charge through doing as much damage as possible, and then move on to the next target. It was quick, daring fighting, and always against impossible odds.

Yet it was effective; the Egyptians suffered many casualties. And as the main force of the operation moved out, Carmi and his men took a much-needed rest. But it was soon interruped. Kibbutz Rvivim was under attack.

Carmi and his men raced off to this isolated outpost in the desert. As they approached, they saw that the kibbutz militia had dug a series of trenches in the sand and were holding off a larger Egyptian force. He ordered his driver to head to the forward line of defense.

Carmi jumped from the still-moving jeep, took up a position in the trench, and began shooting. Only as he reloaded his weapon did he notice the sunburned teenage girl next to him. She was holding a rifle and when he looked at her for another moment, he remembered. They had first met in a church in Poland. "*Shalom*, Eve," he said. But there was no time to talk. The Egyptians were charging across the sand. Carmi and the girl took aim at their enemy, and prepared to fire.

FIFTY-TWO

But this story did not end in the gray sands of the Negev. A more fitting conclusion can be found on a warm spring day almost five decades later. We are in a grassy graveyard outside Ravenna, Italy, and our heroes who were last here as boys have returned as old men, fathers and grandfathers.

There is Pinchuk, a little less hair, heavier, but a prosperous lawyer, chairman of the Israeli War Veterans League. And there is Peltz, as tall and straight-backed in his seventies as when he was a soldier. He became an engineer. In fact, he spent years working on a conveyor system for the Dead Sea Potash Plant. The factory is only a short drive from the spot where he was shot by the Arab bullet that ended his riding. These days, though, he is content to watch his grandchildren ride. They gallop over the fields surrounding the house that he built by the flowering sycamore tree. The brass handle that he had retrieved from Zabiec is fixed to the front door. And there is Carmi. His eyes and hearing are not what they once were, but he appears robust, still broad-shouldered and thick-necked, carrying himself with the authority of a retired army colonel. It is the fiftieth anniversary of the end of the Second World War and they have returned to Ravenna to honor the occasion, and to fulfill one lingering duty.

When they were last in Europe together, the three of them were too busy, too young, and, in truth, too self-involved to appreciate all they had done. Now, from the perspective of old age, they can look back at what the Brigade accomplished and understand its significance. As old men they are sustained by a belief that their youthful dreams have been fulfilled. And more: They are convinced that their deeds helped write history.

For they know that the Brigade demonstrated that Jews could fight back against the Nazis. That they could charge, bayonets high, and win. At a hopeless time, they went into battle as the first Jewish army since the Maccabees, waving a flag with the Star of David.

When it was needed, it was the Brigade that reached out like brothers to the survivors of the Holocaust, offering comfort and strength, and served as emissaries to the Promised Land. Between August 1945 and May 1948, 65 ships carrying 69,878 refugees arrived in Palestine. Another 51,000 arrived from the British detention camps in Cyprus after the State of Israel was established. Many of these immigrants were sent off on this journey by the Brigade.

They were there, too, to defend the new state. As Ben-Gurion wrote, "Without the officers and soldiers of the Jewish Brigade, it is doubtful whether we could have built the Israeli Defense Force in such a short period, in such a stormy hour."

They also helped provide the weapons that allowed the new army to fight back. As well as the troops. The young survivors they had trained in their European camps became twenty thousand combat soldiers in the 1948 war.

But would the new state have come into being, would Israel have survived without the Brigade's fortunate intervention? The answers to such questions, the three old soldiers admit, can only be answered with conjecture.

However, what can be said with more certitude is that it would be a different Israel. And it is this knowledge that fuels their pride. For there is a complex spirit that animates the heart of their small, beleaguered country. It has been woven from many strands: a code of volunteerism and altruism; an attitude of daring and adventure; a commitment to the belief that the Jewish people throughout the world are united in their brotherhood; and a staunchness that proclaims that Jews are no longer passive, easy victims. And these are the sentiments and values that the men of the Brigade through their actions and character helped nurture in the soul of the new nation.

Yet they were young men. They were flawed. They were murderous and vengeful. And perhaps this trait, too, they are willing to concede, has had repercussions.

But the passing years have made them wiser, and they have come to understand the power of a manner of vengeance they could never have imagined in their youth. They have come to Italy in the planes of their own air force, bearing the passports of their own state. It is the very existence of that state, the land they helped create and inspire and defend, that is the ultimate rebuke to the forces who were determined to annihilate the Jews. They have survived and prospered, and that is their vengeance.

So they gather in this country graveyard and slowly walk among the old stones until they find the one they are looking for. A Star of David is carved into the headstone and beneath it is the name HAIM BROT.

Peltz stands in the center, Carmi on one side, Pinchuk on the other, three friends united once again. And together they fulfill one more ancient promise as they chant,

> Yishkadl, veyish kaddash,
> Veyish hadar, veyish haley . . .

And as their memorial prayers continue, they hope that the words of their Kaddish reach up to Heaven; and, their duty done, they can return safely to their own country to live out their days in pride and in peace, and with the satisfaction of knowing that their hard-won treasure of a homeland will be passed on to their children and their children's children, and for all the generations to come.

A NOTE ON SOURCES

This book grew out of a casual, time-killing stroll through a museum. A moment's curiosity as I looked into a glass display case wound up determining the course of the next three years of my life.

It all came about after a Sunday morning trip from my home in Connecticut to the remote, or so it had appeared on the misleading road map I had consulted the night before, reaches of northern New Jersey. As things worked out, the drive was a snap; I arrived at the mammoth community center hours before the scheduled time for my talk about my most recent book, *The Gold of Exodus*. Rather than sit forlornly in my jeep listening to the radio (Casey Kasem's countdown or local church services were the slim early morning pickings), I decided to explore the building. A desultory hour or so later I had already watched a two-on-two pick-up basketball game in the shiny gym and lingered over a cup of coffee in the snack bar when, searching for a new diversion, I went up a flight of stairs and discovered a door marked "HOLOCAUST MEMORIAL MUSEUM."

I entered and quickly realized the title on the door was more impressive than the collection. Lining the walls of a brightly lit space about the size of a motel room were several shoulder-high display cases containing photographs. The black-and-white images—the rage of *Kristallnacht*, a timid doe-eyed boy with a yellow star on his coat being led off by German soldiers—were undoubtedly meant to be an evocative documentation of what had been done a half-century earlier to the Jews of Europe. Yet these photographs were familiar. I had seen identical reproductions in so many books that they had become, at least to me, icons of a sort and as a consequence had lost their power to shock and enrage—and perhaps even a bit of their reality. There are mornings when while sipping my breakfast coffee I see a photo of an emaciated child from the Sudan on the front page of the *Times* and, after just a philosophical there-but-for-the-grace-of-God sigh, go on to check the Yankees box score. I felt a similar detachment as I glanced at the photos in this "museum."

Then something caught my eye. Just as I was leaving, in the glass case by the exit, I noticed a soldier's shoulder patch. A Star of David—the distinctive yellow symbol that victimized Jews had worn on their clothing in photographs throughout the exhibit—was emblazoned on a blue and white field and encircling it were the words, in both Hebrew and English, *Jewish Fighting Brigade.*

Adjacent to the uniform tab, a single sentence had been typed on a three-by-five-inch card: "JEWISH BRIGADE GROUP, a brigade group of the British army, composed of volunteers from Palestine, that was formed in September 1944, and fought in the Italian theater of war from March to May 1945."

A brigade of Jewish soldiers had fought the Nazis as a unit of the British army? I never knew that. What must it have been like, I wondered, to be part of that army, to be a soldier going into combat against an enemy that had set out to destroy your people? What could their war in Europe have been like? These were my initial curiosities, and on the drive home after my talk was finished they still occupied my mind. By the time I had pulled into my driveway, I decided to try and find an answer.

And so I embarked on the investigation that, three years later, resulted in this book. The simplest approach was to read all I could about the Brigade. However, there were only a handful of books on the unit published in English (most notably a thin, disconcertingly organized but still authoritative British volume, *The Jewish Brigade: An Army with Two Masters*, by Morris Beckman; an earnest memoir published in England just after the war by the Brigade's senior chaplain, Bernard M. Casper, *With the Jewish Brigade*; Leonard Sanitt's *On Parade*, a delightful account by a British sergeant-major who had joined the brigade in Italy; Henry Orenstein's fascinating biography of Abram Silberstein, *Abram*; and an artful translation of a powerful and affecting novel originally published in Hebrew three decades ago by a soldier who had served in the unit, Hannoch Bartov, *The Brigade*). The hundreds of documents (daily war diaries, declassified intelligence reports, minutes of cabinet meetings, and ministerial correspondences) obtained from the Public Records Office in London, as well as the archival information gathered at the Central Zionist Archives in Jerusalem, the Beth Hatefutsoth museum, and the War Veterans Committee were, in many ways, more helpful. These papers allowed me to get a broader sense of the political and logistical problem involved in the formation and operation of the Brigade. Additionally, several books and monographs published in Hebrew (Naftali Arbel's *Guns on Senio*; Aharon Hoter-Yishai's *Only Yesterday*; a compilation of articles about the Brigade titled *Collection*; the remarkable and poignant reminiscences of the Doubles edited by Carmy Patale for their 44th reunion, *A Secret Mission*; the War Veterans Committee's "Jewish Palestine Fights Back"; and Yoav Gelbar's thoughtful "Jewish Volunteers in the British Army in World War II") that were translated by my resourceful Tel Aviv–based researcher, Rachel Zetland, served to fill in the historical record in my mind. As further background, I read through a library of histories of World War II and the Holocaust. The most valuable for my purposes were *The Secret Roads*, Jon and David Kimche's postwar British volume about the illegal immigration; Allan Levine's *Fugitives of the Forest*, a wide-ranging account of partisan groups; *The Battle for Italy*, a definitive military history by W. G. F. Jackson; and Lucy S. Dawidowicz's monumental *The War Against the Jews*.

But I quickly realized that to get an answer to my initial questions

I had to get another kind of "data." I needed to speak to the men who had served in the Brigade. Fortunately, many of these soldiers, now of course old men, were available for interviews by either myself or my research associates. (A list of the veterans interviewed is included at the end of this note.) In Israel I sat on a screened porch overlooking a plowed field in a kibbutz near Haifa, toured a shooting range in Carmel with the steady *ratatat* of Uzis ringing in my ears, sipped tea in a sparklingly white kitchen on a hilltop in Carmel, a view of the sun-dappled Mediterranean in the background—and all the while listened as stories about the war in Europe and its aftermath were told. I went on a fascinating and moving intellectual journey to homes throughout the small country. And as I spoke with these men I felt the growing excitement of one who had entered a rich and unexplored, in a literary sense, terrain: the secret, or at least unreported, drama of the soldiers who had served in the Brigade.

As a result, a different sort of book from the one I had originally imagined began to take shape in my mind. I realized that if I were to tell this story accurately, if I were to do justice to both the recorded facts and the psychological mysteries, I would need to write a narrative that wove together both of these strands. I would have to do the historian's job of trying to understand and describe what happened. And I would also need to go beyond the documents to give the reader a sense of what the people in this story were thinking and feeling as their lives were radically transformed. Another challenge: It had to be true.

Of course, such a narrative strategy is filled with problems. Oral histories—whether interviews or memoirs—can be subjective. People's memories (as well as prejudices and egos) can slant events, get facts wrong, or even be intentionally misleading. Nevertheless, in a saga like the one I was setting out to tell, archival documents, the historian's traditional standard for accuracy, are despite their seeming authority often incomplete. For example, the war diary of the Brigade, obtained from the British Military Archives, Army Form C. 2118, reads as follows for November 1, 1944: "Bde at sea." True—

as far as it goes. But as I recount in Part I (the first three chapters) of this book, an intriguing drama was taking place in the hearts and minds of these soldiers while the "Bde at sea."

As my research progressed, I became convinced that the most effective way to tell this story would be to narrow my focus. I wanted to write a book that was representative of the experiences of a brigade of men, yet I would tell it through a handful of protagonists. But there were five thousand soldiers. Whom should I focus on? As the reader by now is well aware, I zeroed in on three friends—Israel Carmi, Johanan Peltz and Arie Pinchuk. I will let the reader judge if what attracted me to these men was worthy of the attention I directed on their histories. However, it should be noted that not only their lives were remarkable. Each of my "heroes" (I confess: I grew to like and respect these men immensely) came to me bearing unplundered literary treasures. Carmi had written an autobiographical account in Hebrew of his military service for the years leading up to and during World War II, *The Fighter's Path*. I had Ms. Zetland translate it, and what I read was a page-turning personal history, one that was never sentimental, but often provocative. Peltz's autobiography is still a massive work-in-progress. But the completed sections he kindly shared were written in a carefully crafted English and provided a detailed and absorbing story of his life at Zabiec, in Palestine, and at war. He is a natural writer, his tone by turns amusing, poignant—and always compelling. Pinchuk shared—and Ms. Zetland translated—a privately published memorial book about Reflovka, a 470-page heartfelt volume of articulate reminiscences from him, his sister, and others. In addition to their books and manuscripts, these three men and Leah Pinchuk Zeiger were generous enough to sit through hours of interviews, then more interviews, and they continued to respond graciously to the frequent small questions I had once I began writing and editing the book.

So in the end, I had a wealth of information to draw on. And I was determined to use all these varied sources—archival, formal histories, memoirs, and interviews—in order to craft a true story. Therefore, when there are direct quotations the reader can be assured that at least one of the participants in the conversation remembers what was said exactly as recorded. And when there is an extended dialogue, an exchange that would be asking a great deal of anyone to remember verbatim after fifty years, this, too, is not arbitrary. For example, on the bus tide to Rome (chapter eight) when Peltz tells the story of his fights in Hartuv, I am quoting his words directly as they appeared in his autobiography.

Further, although I write about the attack on La Giorgetta largely from Peltz's point of view, I used a variety of sources both to confirm his memory and to re-create the drama: Peltz's memoir, the war diary documents, declassified intelligence reports, Jackson's history of the war in Italy, several books about mines and the sapper battalions (most prominently *War in the Desert* by James Lucas), interviews with several other participants including Peltz's sergeant Avraham Uzieli, Beckman's *The Jewish Brigade*, and the transcript of an interview with the Israeli historian Joav Gelber. Another example: When I write about Leah's experience—what she saw, felt, and thought—the reader should understand that these sections do not simply rely solely on her memories, but have been confirmed and expanded with the help of more formal histories such as Levine's *Fugitives of the Forest*, and the written reminiscences of others who lived in Reflovka including, most significantly, Ziskin-Haim Bert, Rachel Fligelman, and Ahuva Hendelsman.

I also found valuable information, it would be remiss not to acknowledge, in another resource: filmed documentaries. An Israeli television documentary, *Hanokmim (The Avengers),* and a tape of a shockingly candid talk Meir Zorea gave at his kibbutz that his widow kindly provided (both translated for me by Ms. Zetland) offered incredible eyewitness testimony about the pursuit of vengeance. For example, when Zorea discusses his technique for executing war criminals (chapter thirty-four), this account is taken directly from the talk he gave to the members of his kibbutz shortly before his death. No less valuable was an insightful and extremely evocative film history of the Brigade directed by Chuck Olin, *In Our Own Hands*. The producers—Olin, Chuck Copper, and Matthew Palm—

have a website, www.olinfilms.com, that any reader interested in this book would want to consult.

All these sources, then, allowed me to attempt to answer the questions that first rose up in my imagination on that fateful morning in New Jersey. They are the factual underpinnings that support my intent to tell this story with drama, as well as with insight and authority.

Interviews with the following veterans were invaluable, and their reminiscences helped animate the entire book:

Abraham Akavia
Arie Amir
Ted Arison
Ephraim Ben Arzi
Eli Avni
Izhak Bar-On
Yehezkel Bar-On
Hannoch Bartov
Zve Brenner
David Ben-David
Gideon Ben Israel
Israel Carmi
Eric Feuer
Oly Givon
Cyril Goodman
Chanan Greenwald
Martin Hauser
Aharon Hoter-Ishay
Mark Hyatt
Maxim Kahan
Gabriel Knoller
Jack Levy
Netanel Lorch
Shimon Maze
Pavel Mozes
Shlomo Netzer (a survivor who became an "honorary" member of the Brigade)
Johanan Peltz
Arie Pinchuk
Shaul Ramati
Joseph Seltzer
Meir de Shalit
Aharon Shamir
Shlomo Shamir
Abram Silberstein
Israel Tal
Adin Talbar
Avraham Uzieli
Meir Zorea

The primary sources for each chapter in this book are as follows:

PROLOGUE

Documents: War Cabinet Memorandum, "Palestinians in the Forces" (8/1/42); War Office Telegrams, 1940–43; Cabinet and Colonial Office correspondence, 1940–44; Army

Council Secretariat Report, 10/42; Prime Minister's correspondence, 1940–44; Colonial Office telegrams, 1940–44; Middle Eastern War Council Report, 5/24/42; Secretary of State Brief on "Jewish Army," 7/13/42; Weizmann–Churchill correspondence, 1942–44.

Monographs: Jewish Brigade Group recruiting brochure, published by Jewish Agency for Palestine; Churchill's House of Commons speech, 9/28/44, reprinted by Jewish Agency; Martin Gilbert, "Churchill and Zionism" (The World Jewish Congress); David Ben-Gurion, "To a Comrade in Palmach" (Ministry of Defense); Joav Gelber, "Jewish Volunteers in the British Army in WW II."

Newspapers: *New York Times*, Sept.–Oct. 1944; Central Zionist Palestinian press archives 1939–1945.

Books: Morris Beckman, *The Jewish Brigade* (Spellmount); Henry Orenstein, *Abram* (Beaufort Books); Chaim Herzog, *The Arab-Israeli Wars* (Vintage); Yehuda Bauer, *From Diplomacy to Resistance* (Jewish Publication Society of America); Michael Cohen, editor, *The Jewish Military Effort, 1939–44* (Garland); Jehuda Wallace, *Israeli Military History* (Garland).

Interviews: Brigade veterans; Yoav Gelber (YG).

CHAPTER ONE

Documents: Brigade War Diary; Brigade HQ memorandum.

Books: Bernard Casper, *With the Jewish Brigade* (Edward Goldston); Beckman; Sam Axelrod, *My Story* (Midax Press); Israel Carmi, *The Fighter's Path* (Ma'arakhot; Zetland translation).

Interviews: Israel Carmi (IC); Johanan Peltz (JP); Arie Pinchuk (AP); Brigade veterans.

Memoirs: Johanan Peltz, "With the Palestinian Police."

CHAPTER TWO

Interviews: IC; JP; AP; Brigade veterans

Memoirs: Peltz; Arie Pinchuk, "Reflovka."

Books: Carmi; Bauer; Herzog; Beckman.

CHAPTER THREE

Interviews: AP; JP; IC.

Memoirs: Peltz, "My Big Grandfather," "The Furriers"; Pinchuk.

Books: Beckman; Casper.

CHAPTER FOUR

Documents: War Diary; Intelligence Reports; Brigade HQ memorandum.

Books: Beckman; Axelrod; W. G. F. Jackson, *The Battle for Italy* (Harper & Row); Herzog; Bauer; Cohen; Ros Belford et al., *The Real Guide: Italy* (Prentice Hall); Carmi.

Interviews: IC; Brigade veterans.

CHAPTER FIVE

Books: Carmi; Bauer; Cohen; Anthony Cave Brown, *Bodyguard of Lies* (Harper & Row).

Interviews: IC; JP; Oly Givon (OG).

CHAPTER SIX

Documents: War Diary; Brigade HQ memorandum.
Memoirs: Peltz.
Books: Beckman; Casper.

CHAPTER SEVEN

Books: Beckman; Casper; Axelrod; Leonard Sanitt, *On Parade* (Spa Books).
Memoirs: Pinchuk.
Interview; AP.

CHAPTER EIGHT

Books: Beckman; Axelrod.
Interviews: AP; JP; IC.
Memoirs: Peltz; Pinchuk.

CHAPTER NINE

Memoirs: Leah Pinchuk Ziegler; AP; Ziskin-Haim Bert; Ahuva Hendelsman; Gershon Gruber; Itzhak Bril; Rivka Avira; Enya Burku; Tziporah Goldberg and others in "Reflovka."
Books: Lucy S. Dawidowicz, *The War Against the Jews* (Holt, Rinehart and Winston).
Interviews: Leah Pinchuk Zieger (LPZ); AP.

CHAPTER TEN

Documents: War Diary; Intelligence Reports.
Interviews: AP; JP; IC; Brigade veterans.
Books: Jackson; Beckman; Casper; Axelrod; Cave Brown; Carmi.

CHAPTER ELEVEN

Documents: War Diary; Intelligence Reports; Brigade HQ memorandum.
Interviews: JP; YG; Avraham Uzieli (AU); Izhak Bar-On (IB); Shlomo Shamir (SS); Maxim Kahan (MK); Zve Brenner (ZB); Hannoch Bartov (HB); Brigade veterans.
Memoirs: JP.

CHAPTER TWELVE

Documents: War Diary; Intelligence Reports.
Interviews: JP; MK; Brigade veterans.
Memoirs: JP.
Documentary: *In Our Own Hands.*

CHAPTER THIRTEEN

Memoirs: LPZ; others in "Reflovka."
Interviews: LPZ; AP.
Books: Dawidowicz.

CHAPTER FOURTEEN

Documents: War Diary; Intelligence Reports.

Books: Beckman; James Lucas, *War in the Desert* (Holt); Jackson.

Interviews: JP; AU; MK; Brigade veterans.

CHAPTER FIFTEEN

Documents: War Diary; Military Cross Citation for Meir Zorea.

Interviews: JP; AU; MK; IC; Meir Zorea (MZ); AP; Brigade veterans.

Books: Beckman; Jackson.

Memoirs: JP.

CHAPTER SIXTEEN

Documents: War Diary.

Interviews: JP; AU; MK; HB; Brigade veterans.

Books: Beckland.

Memoirs: Peltz.

Documentary: *In Our Own Hands.*

CHAPTER SEVENTEEN

Documents: War Diary; Intelligence Reports.

Interviews: IC; Brigade veterans.

Books: Carmi; Beckman.

CHAPTER EIGHTEEN

Documents: War Diary; Intelligence Reports; Brigade HQ memorandum.

Interviews: JP; IC; MZ; AU; SS; IB; ZB; AP; Brigade veterans.

Books: Carmi; Beckland; Taylor; Hannoch Bartov, *The Brigade* (Macdonald).

CHAPTER NINETEEN

Interviews: LPZ

Memoirs: LPZ; others in "Reflovka"

CHAPTER TWENTY

Documents: War Diary.

Books: Axelrod; Casper; Sanitt; Beckland; Carmi.

Interviews: JP; IC; AP; SS; MK; Brigade veterans.

CHAPTER TWENTY-ONE

Interviews: LPZ; AP.

Memoirs: LPZ; AP.

Books: Allan Levine, *Fugitives of the Forest* (Stoddart); Raul Hilberg, ed., *Documents of Destruction* (Quadrangle); Gerd Korman, ed., *Hunters and Hunted* (Delta).

CHAPTER TWENTY-TWO

Documents: War Diary.

Books: Carmi; Beckland; Casper; Peter Rabe, *Palmach's German Platoon in the Western Desert* (Amihai).

Interviews: IC; SS; Chanan Greenwald (CG); HB; JP; OG; Brigade veterans.

CHAPTER TWENTY-THREE

Documents: War Diary; Intelligence Reports.

Books: Carmi; Jackson; Beckland; Axelrod; Casper.

Interviews: MK; JP; AU; IC; SS; AP; Brigade veterans.

CHAPTER TWENTY-FOUR

Interviews: LPZ.

Memoirs: LPZ.

Books: Levine; Korman; Hilberg

CHAPTER TWENTY-FIVE

Document: War Diary; Brigade HQ memorandum.

Books: Bartov; Beckland; Axelrod; Casper.

Interviews: IC; JP; AP; SS; HB; Brigade veterans.

Documentary: *In Our Hands.*

CHAPTER TWENTY-SIX

Documents: War Diary; Brigade HQ memorandum.

Books: Bartov; Beckland; Axelrod; Carmi; Casper.

Interviews: IC; JP; AP; Brigade veterans.

CHAPTER TWENTY-SEVEN

Books: Beckland; Casper; Carmi; Hilberg; Dawidowicz.

Interviews: IC; JP; Hoter-Yishai (HY).

Memoirs: JP.

CHAPTER TWENTY-EIGHT

Interviews: LPZ

Memoirs: LPZ; others in "Reflovka."

CHAPTER TWENTY-NINE

Documents: War Diary; Intelligence Reports.

Books: Beckland; Carmi; Bartov.

Interviews: IC; SS; JP; OG; MZ; Abram Silberstein (AS); Brigade veterans.

Newspaper: Benny Morris, "Interview/Israel Carmi," *The Jerusalem Post Magazine*, 7/7/89.

Documentaries: *Hanokmim*, Israeli television; *In Our Hands*; Zorea tape.

CHAPTER THIRTY

Book: Carmi.

Interview: IC; OG.

Documentary: *Hanokmim*

CHAPTER THIRTY-ONE

Interviews: LPZ

Memoirs: LPZ.

CHAPTER THIRTY-TWO

Interviews: JP; IC; MZ; AS; OG; AP.

Book: Carmi; Beckland; Axelrod.

Newspaper: *Jerusalem Post* profile of Carmi.

Documentaries: *Hanokmim*; Zorea tape.

CHAPTER THIRTY-THREE

Interviews: JP; IC; OG.

Books: Rabe.

Documentaries: *Hanokmim*; *In Our Own Hands.*

CHAPTER THIRTY-FOUR

Interviews: IC; JP; MZ; AS; OG.

Documentaries: *Hanokmim*; Zorea tape.

Books: Carmi; Beckland.

CHAPTER THIRTY-FIVE

Interviews: JP; IC; AS; OG.

Memoirs: Peltz.

Documentary: *Hanokmim.*

CHAPTER THIRTY-SIX

Interviews: JP; IC; AP.

CHAPTER THIRTY-SEVEN

Documents: War Office Intelligence Reports; War Diary; Zionist Archive "Special Reports" on refugees.

Books: Carmi; Casper; Jon and David Kimche, *The Secret Road* (Secker and Warburg); Beckland; Aharon Hoter-Yishai, *Only Yesterday* (Ma'arakhot); Dominique Lapierre, *A Thousand Suns* (Warner); Dawidowicz.

Monograph: "Studies in the History of Zionism," The Pedagogic Center, Dr. Motti Friedman, director.

Interviews: IC; JP; SS; CG; HY; Brigade veterans.

CHAPTER THIRTY-EIGHT

Documents: "Special Reports" on Refugees; British Army Field Service Intelligence Reports.

Interviews: AP.

Memoirs: AP.

Books: Kimche; Hoter-Yishai.

CHAPTER THIRTY-NINE

Document: British Army Field Service Intelligence Reports.

Books: Carmi; Beckland; Sanitt; Axelrod.

Interviews: IC; JP; SS; HB; MK; Brigade veterans.

Memoirs: Lisa and Aaron Derman; Yonathan Adar, "Bricha Route Through the Alps."

Documentary: *In Our Own Hands.*

CHAPTER FORTY

Books: Carmi; Kimche.

Interviews: IC; JP; Brigade veterans.

CHAPTER FORTY-ONE

Books: Carmi; Kimchel; Orenstein.

Interviews: IC; JP; MK; SS; AS; Meir de Shalit (MS); Brigade veterans.

CHAPTER FORTY-TWO

Books: Carmi, Kimche.

Interviews: IC; JP; MS; MK; AS; Brigade veterans; Lisa and Aaron Derman.

CHAPTER FORTY-THREE

Memoirs: Pinchuk.

Interviews: AP; LPZ.

CHAPTER FORTY-FOUR

Documents: Brigade HQ memorandum; War Diary.

Books: Carmi; Bartov; Beckland; Casper; Axelrod; Herzog; Kimche; I. F. Stone, *Underground to Palestine* (Boni & Gaer).

Interviews: IC; JP; SS; HB; OG; MZ; MS; Brigade veterans.

Memoirs: Ze'hava Litman Brumberg, "School in Bergen-Belsen Revived My Soul"; Netanael Lorch, "The Immigration of a Double."

CHAPTER FORTY-FIVE

Memoirs: Pinchuk.

Interviews: AP; LPZ.

CHAPTER FORTY-SIX

Books: Carmi; Kimche; Stone; Lapierre.

Interviews: IC; JP; Brigade veterans.

CHAPTER FORTY-SEVEN

Documents: Beth Hatefutsoth "Return to Life" exhibition and catalog; Zionist Archives.

Books: Carmi; Kimche; Stone; Lapierre; Bauer; *Encyclopedia Judaica* (Macmillan)

Interviews: IC; JP; Gelber.

CHAPTER FORTY-EIGHT

Interviews: AP; LPZ.

Memoirs: AP; LPZ.

Newspaper: Joseph Berger, "Grasping Life After War," *The New York Times*, 1/15/00.

CHAPTER FORTY-NINE

Interviews: IC; JP; AP; HB; SS; Gabby Knoller; Brigade veterans.

Memoirs: AP; Gabby Knoller, "Leni."

Book: Carmi; Beckman.

CHAPTER FIFTY

Documents: War Diary; Colonial Office telegrams; Zionist Archive papers; Beth Hatefutsoth collection.

Books: Carmi; Orenstein; Beckman; Kimche; Herzog; Lapierre; Stone; Casper; Beckland.

Memoir: Carmy Patael, editor, "The Doubles in a Secret Mission" (translated by Zetland).

Interviews: IC; JP; AP; SS; Brigade veterans.

CHAPTER FIFTY-ONE

Books: Carmi; Orenstein; Herzog.

Interviews: AP; LPZ; IC; JP.

CHAPTER FIFTY-TWO

Documents: Beth Hatefutsoth collection.

Interviews: IC; JP; AP; SS; HB; JG.

Books: Kimche; Beckman.

ACKNOWLEDGMENTS

Although sequestered in my hideout two floors above a Main Street pizza parlor in a small New England town for a seemingly endless stretch as I wrote this book, I was never really on my own. From the moment I shared my still inchoate idea with my agent (and friend) Lynn Nesbit, she offered shrewd and valuable support. Cullen Stanley, of the the Janklow-Nesbit agency, was also as always full of wisdom, kindness, and friendship throughout the entire process. While at HarperCollins, I was fortunate to be in the hands of a skilled, insightful, and—no less important an attribute when invariably things get complicated—straight-shooting editor, David Hirshey. Jeff Kellogg, also at HarperCollins, proved to be a deft and incisive editor; this book is in many ways a product of his attention and talent. Further, Brenda Segel and Patricia Kelly were also generous with their help and enthusiasm. And even before there was a manuscript, Graydon Carter, Doug Stumpf, and Chris Garrett at *Vanity Fair* were full of support; I'm grateful for the research trip they sponsored to Israel. Similarly, Nick Wechsler of Industry Entertainment and Harvey Weinstein and Jon Gordon of Miramax Films became involved in making a movie from this book even before there was a book; their energy and interest were valued companions as I wrote on. And it was the always resourceful and erudite

Bob Bookman who made the initial introductions. Alan Hergott not only made sure all the *i*'s were dotted and the *t*'s crossed, but also was a constant source of wisdom and friendship; I'm in his debt. Working with me in Israel, Rachel Zetland was an untiring guide, artful translator, and probing interviewer; without her help for the past three years, I doubt this book could have been written. Also in Israel, I'm grateful to the many people who were kind enough to allow me into their homes and who shared the stories of their lives with me. There are too many people to thank by name; however, I would be remiss if I didn't formally express my gratitude to Johanan Peltz, Israel Carmi, Arie Pinchuk, and Leah Pinchuk Ziegler. My life was enriched by the time I spent with them. Finally, there's my family—my wife, Jenny, and my children, Tony, Anna, and Dani. I wrote this book for them.